Superhydrophobic Coatings for Corrosion and Tribology

Superhydrophobic Coatings for Corrosion and Tribology

Special Issue Editors

Shuncai Wang
Guochen Zhao

MDPI • Basel • Beijing • Wuhan • Barcelona • Belgrade

MDPI

Special Issue Editors

Shuncai Wang
University of Southampton
UK

Guochen Zhao
Qilu University of Technology
(Shandong Academy of Sciences)
China

Editorial Office
MDPI
St. Alban-Anlage 66
4052 Basel, Switzerland

This is a reprint of articles from the Special Issue published online in the open access journal *Coatings* (ISSN 2079-6412) from 2018 to 2019 (available at: https://www.mdpi.com/journal/coatings/special_issues/superhydrophobic_coat_corrosion_tribology)

For citation purposes, cite each article independently as indicated on the article page online and as indicated below:

LastName, A.A.; LastName, B.B.; LastName, C.C. Article Title. *Journal Name* **Year**, *Article Number*, Page Range.

ISBN 978-3-03921-784-7 (Pbk)
ISBN 978-3-03921-785-4 (PDF)

Contents

About the Special Issue Editors

Shuncai Wang (Dr.) is the academic lead at the University Science and Engineering Electron Microscopy Centre, Lecturer in the National Centre for Advanced Tribology at Southampton (nCATS) at the University of Southampton and Fellow of Royal Microscopical Society. He obtained his Ph.D. in 1998 at the School of Metallurgy and Materials, University of Birmingham. He won a number of grants including the Innovation China-UK Knowledge transfer project (2008–2010); Royal Society Visiting Fellowship (2008) and International Exchange Programme (2012–2014 and 2016–2018); Royal Academy of Engineering Research Exchange (2014–2015) and Distinguished Visiting Fellowship (2016); and EPSRC—Global Challenge Research Fund (2016–2017). He also works closely with the industry, including Boeing, Airbus, GKN Aerospace, TWI, Daido Metals. He won the RCUK knowledge transfer twice and organised the following events of the UK-China Symposium in Tribology (2009 in Lanzhou, 2010 in Southampton), which is now an annual joint event between UK and China Tribology Society. He has supervised 13 Ph.D. and 60 MSc/MEng/BEng and published over 120 peer-viewed journal papers with an h-index of 32, i10-index 62, and citations over 3600 by electron microscopical characterization and fabrication of engineering materials and surface coatings by vapour deposition, thermal spraying and electrodeposition. His current research interests are in gradient, low frictional, wear-resistant, biomimetic, and superhydrophobic coatings.

Guochen Zhao (Dr.) is a research fellow in Shandong Provincial Key Laboratory of High Strength Lightweight Metallic Materials, Advanced Materials Institute, Qilu University of Technology (Shandong Academy of Sciences), China. He received his Ph.D. from Shanghai University in Dec. 2016 and joined the key laboratory in Aug. 2017. He took part in Erasmus programme for 6 months researching at the University of Chemical Technology and Metallurgy—Sofia (2012–2013) and spent one year at the University of Southampton as an academic visitor (2014–2015). His research interests are superwettability properties, robust superhydrophobic coatings, antifouling anticorrosion films, superaerophilic electrodes, and 3D printing. He has published over 20 peer-reviewed journal papers in *Coatings, Appl. Surf. Sci., Surf. Coat. Technol., RSC Adv.,* etc., and owns six P.R.C. patents.

MDPI

Editorial

Special Issue on Superhydrophobic Coatings for Corrosion and Tribology

Shuncai Wang [1],* and Guochen Zhao [2],*

[1] National Centre for Advanced Tribology at Southampton (nCATS), University of Southampton, Southampton SO17 1BJ, UK

[2] Shandong Provincial Key Laboratory of High Strength Lightweight Metallic Materials, Advanced Materials Institute, Qilu University of Technology (Shandong Academy of Sciences), Jinan 250014, China

* Correspondence: wangs@soton.ac.uk (S.W.); zhaogch@sdas.org (G.Z.)

Received: 9 October 2019; Accepted: 12 October 2019; Published: 14 October 2019

1. Introduction

Superhydrophobicity, showing strong water-repellency, has been widely investigated for many applications, especially in the fields of corrosion protection and antifouling. Water tends to roll off from superhydrophobic surfaces like natural lotus leaves. When a corrosive aqueous solution comes into contact with such a surface, a stable air cushion is formed on the interface between liquid and solid which minimizes the contact area. As a result, the charge transfer of the corrosive reaction is dramatically restrained, resulting in a positively shifted corrosion potential and low corrosion rate. Additionally, the superhydrophobic surface effectively isolates microorganisms from adhering on the surface and thus prevents microbiologically influenced corrosion caused by their metabolites. Thus, the superhydrophobic coatings have potential applications in corrosion protection of marine equipment, medical devices, mechanical components, etc.

However, the lack of mechanical strength and heat resistance prevents the use of these coatings in harsh environments. It is well established that micro-nano hierarchical structures and low surface energy are the two fundamental factors crucial to developing superhydrophobic surfaces, and the superhydrophobicity of these surfaces would be diminished if they were destroyed by abrasion or overheating. The superhydrophobic coatings using wear-resistant inorganic materials are therefore highly sought after. Ceramics are of particular interest due to their high mechanical strength, heat and corrosion resistance. Such superhydrophobic coatings have recently been successfully fabricated using a variety of ceramics and different approaches, and have shown improved wear and tribocorrosion resistance properties.

This special issue is making the best effort to reflect the recent developments in the fabrication of superhydrophobic coatings and their robustness against corrosion and wear resistance. We hope it will stimulate the future research and application.

2. Superhydrophobic Coatings for Corrosion and Tribology

The special issue "Superhydrophobic Coatings for Corrosion and Tribology" was opened in June 2018 and closed in June 2019. During this period of 12 months, there were 19 manuscripts submitted to this special issue, of which 10 were finally accepted. Among the publications, most works focused on the fabrication of superhydrophobic surfaces and applications of corrosion protection, self-cleaning, and oil-water separation. The tribological coatings for corrosion and wear resistance were also included to facilitate other extended functions. The individual work is briefed below:

The paper "Fabrication of Superhydrophobic AA5052 Aluminum Alloy Surface with Improved Corrosion Resistance and Self Cleaning Property" by Zhao et al. presents a method for fabricating superhydrophobic Ni–Co alloy coatings on aluminum surfaces, followed by surface modification with 6-(*N*-allyl-1,1,2,2-tetrahydro-perfluorodecyl) amino-1,3,5-triazine-2,4-dithiol monosodium (AF17N) [1]. The authors discussed the importance of grooves to superhydrophobicity and the effects of AF17N polymeric nanofilm on the corrosion inhibition property using cyclic voltammetry (CV) curves. The coating showed excellent self-cleaning and corrosion-resistance performance with good chemical stability and long-term durability of more than 16 weeks, which is essential for practical applications.

The paper "Effect of Surface Topography and Structural Parameters on the Lubrication Performance of a Water-Lubricated Bearing: Theoretical and Experimental Study" by Xie et al. described theoretical calculations and experiments examining the influence of the surface topography on water-lubricated bearings' lubrication performance [2]. The test data and simulation corresponded well, showing the influences of bushing, eccentricity ratio, bushing deformation, specific pressure, speed, velocity, etc. on the lubrication performance. This work elucidates the lubrication mechanism and is of significance for designing such bearings.

The paper "Effects of Surface Microstructures on Superhydrophobic Properties and Oil-Water Separation Efficiency" by Chen et al. was devoted to increasing the efficiency of oil-water separation by fabricating superhydrophobic copper meshes [3]. Regarding the effects of microstructures of membranes, they found meshes with parabolic morphology to have better separation efficiency than the truncated cone morphology, up to 97.5%, 97.2% and 91% for benzene-water, carbon tetrachloride-water and engine oil-water mixtures, respectively. This work clarified the effects of microstructures relevant to superhydrophobicity which may guide the applications.

The paper "Surfactant-Free Electroless Codeposition of Ni–P–MoS$_2$/Al$_2$O$_3$ Composite Coatings" by Liu et al. presents the tribological performance of electroless surfactant-free Ni–P–MoS$_2$/Al$_2$O$_3$ coatings [4]. The incorporation of Al$_2$O$_3$ coated MoS$_2$ particles obviously improved the wear resistance of such coatings. In the absence of surfactants, compact Ni–P–MoS$_2$/Al$_2$O$_3$ coatings with 7% Al$_2$O$_3$ were obtained, which had a low friction coefficient of 0.4 and lower mass loss of wear than Ni–P–MoS$_2$ coatings.

The paper "Synthesis and Properties of Electrodeposited Ni–Co/WS$_2$ Nanocomposite Coatings" by He et al focused on the Ni–Co/WS$_2$ superhydrophobic coatings [5]. The one-pot plated Ni–Co coating had an ultra-low friction coefficient of 0.16 with the embedment of 7.1 wt.% WS$_2$ lubricants, while also showing excellent superhydrophobicity with a water contact angle of 157°. This work successfully prepared a superhydrophobic surface which possessed self-lubrication, high abrasive resistance and good mechanical properties.

The paper "Super-Hydrophobic Co–Ni Coating with High Abrasion Resistance Prepared by Electrodeposition" by Xue et al. demonstrated the fabrication of Co–Ni superhydrophobic coatings on carbon steel with excellent abrasive resistance and anti-corrosion properties [6]. In their work, the measurement of cyclic voltammograms contributed to determining the appropriate potential for constructing the hierarchical structures of Co–Ni coatings. Modified by 1H,1H,2H,2H-Perfluorooctyltrichlorosilane (PFTEOS), the coating acquired superhydrophobicity with contact angles over 161° and could maintain this kind of property after a 12 m abrasion test. The trapped air on the surface of the coating provided stable corrosion resistance with corrosion rate more than 20 times lower than bare carbon steel.

The paper "Oscillating Magnetic Drop: How to Grade" authored by Goncalves Dos Santos et al. described an alternative method to goniometer-based measurements for evaluating superhydrophobicity: the measurement of the damped-oscillatory motion of a ferrofluid sessile droplet [7]. Through comparing results measured by the two methods, it was found that the damping time of oscillating magnetic drops had a more sensitive response than the traditional one. This method explored by the authors had great significance to grading superhydrophobic surfaces.

The paper "Robust Super-Hydrophobic Coating Prepared by Electrochemical Surface Engineering for Corrosion Protection" by Bi et al. reviewed the recent advances in fabricating robust superhydrophobic surfaces by electrochemical methods, including electrochemical anodization, micro-arc oxidation, electrochemical etching and electrochemical deposition and their corrosion resistance [8]. The wetting principle, corrosion protection mechanism and stability of superhydrophobic coatings were also described in detail, inspiring the fabrication of promising multi-functional coatings for self-healing, slippery liquid-infused porous surface etc.

The paper "Preparation of Superhydrophobic Steel Surfaces with Chemical Stability and Corrosion" by Du et al. presents a simple method for fabricating superhydrophobic surfaces on Q235 steel [9]. The needle-like structures followed by fluorination treatment process exhibited extremely low water affinity (with a water-contact angle of 161°) and high elastic bounce back, resulting in a low-corrosion current which represents about 8.7% of the untreated one. Moreover, such superhydrophobicity could be maintained in various aqueous solutions with different pH values.

The paper "Structure-Property Relationships in Suspension HVOF Nano-TiO$_2$ Coatings" authored by Zhang et al. shows the effects of microstructure on mechanical properties and wear behavior [10]. By controlling thermal spraying parameters, the stable suspension leads to the formation of quality high-velocity oxygen fuel (HVOF) TiO$_2$ coatings with the hardness as high as 7.8 GPa and low coefficient of friction and wear rate as low as 0.35 and 0.83 mm^3/Nm, respectively. This work provided good guidance for depositing thermally-sensitive materials such as TiO$_2$-anatase or hydroxyapatite, with minimal thermal decomposition.

3. Perspectives

Superhydrophobic coatings have attracted extensive attention. Further research may focus on improving the surface stability and robustness. The multiple functional coatings, as published in this special issue, are also highly sought after to achieve both superhydrophobicity and tribocorrosion. A combination of superhydrophobicity and self-lubricated surfaces are another case to broaden their applications in fields of energy, biology, sensor, environment, etc.

Acknowledgments: We would like to show our appreciation to all the authors, reviewers and editors for their contribution in this special issue of *Coatings*.

Conflicts of Interest: The authors declare no conflict of interest.

References

1. Zhao, Q.; Tang, T.; Wang, F. Fabrication of Superhydrophobic AA5052 Aluminum Alloy Surface with Improved Corrosion Resistance and Self Cleaning Property. *Coatings* **2018**, *8*, 390. [CrossRef]
2. Xie, Z.; Rao, Z.; Liu, H. Effect of Surface Topography and Structural Parameters on the Lubrication Performance of a Water-Lubricated Bearing: Theoretical and Experimental Study. *Coatings* **2019**, *9*, 23. [CrossRef]
3. Chen, Y.; Yang, S.; Zhang, Q.; Zhang, D.; Yang, C.; Wang, Z.; Wang, R.; Song, R.; Wang, W.; Zhao, Y. Effects of Surface Microstructures on Superhydrophobic Properties and Oil-Water Separation Efficiency. *Coatings* **2019**, *9*, 69. [CrossRef]
4. Liu, P.; Zhu, Y.; Shen, Q.; Jin, M.; Zhong, G.; Hou, Z.; Zhao, X.; Wang, S.; Yang, S. Surfactant-Free Electroless Codeposition of Ni–P–MoS$_2$/Al$_2$O$_3$ Composite Coatings. *Coatings* **2019**, *9*, 116. [CrossRef]
5. He, Y.; Wang, S.; Sun, W.; Reed, P.A.S.; Walsh, F.C. Synthesis and Properties of Electrodeposited Ni–Co/WS$_2$ Nanocomposite Coatings. *Coatings* **2019**, *9*, 148. [CrossRef]
6. Xue, Y.; Wang, S.; Bi, P.; Zhao, G.; Jin, Y. Super-Hydrophobic Co–Ni Coating with High Abrasion Resistance Prepared by Electrodeposition. *Coatings* **2019**, *9*, 232. [CrossRef]
7. Goncalves Dos Santos, A.; Montes-Ruiz Cabello, F.J.; Vereda, F.; Cabrerizo-Vilchez, M.A.; Rodriguez-Valverde, M.A. Oscillating Magnetic Drop: How to Grade Water-Repellent Surfaces. *Coatings* **2019**, *9*, 270. [CrossRef]
8. Bi, P.; Li, H.; Zhao, G.; Ran, M.; Cao, L.; Guo, H.; Xue, Y. Robust Super-Hydrophobic Coating Prepared by Electrochemical Surface Engineering for Corrosion Protection. *Coatings* **2019**, *9*, 452. [CrossRef]

9.   Du, C.; He, X.; Tian, F.; Bai, X.; Yuan, C. Preparation of Superhydrophobic Steel Surfaces with Chemical Stability and Corrosion. *Coatings* **2019**, *9*, 398. [CrossRef]

10.   Zhang, F.; Wang, S.; Robinson, B.W.; Lovelock, H.L.d.V.; Wood, R.J.K. Structure-Property Relationships in Suspension HVOF Nano-TiO$_2$ Coatings. *Coatings* **2019**, *9*, 504. [CrossRef]

coatings

MDPI

Review

Robust Super-Hydrophobic Coating Prepared by Electrochemical Surface Engineering for Corrosion Protection

Peng Bi [1,†], Hongliang Li [1,†], Guochen Zhao [2,*], Minrui Ran [3], Lili Cao [4], Hanjie Guo [1,*] and Yanpeng Xue [3,*]

1   School of Metallurgical and Ecological Engineering, University of Science and Technology Beijing, Xueyuan Road 30, Beijing 100083, China
2   Shandong Provincial Key Laboratory for High Strength Lightweight Metallic Materials, Advanced Materials Institute, Qilu University of Technology (Shandong Academy of Sciences), Jinan 250000, China
3   National Center for Materials Service Safety, University of Science and Technology Beijing, Xueyuan Road 30, Beijing 100083, China
4   School of Mechanical and Energy Engineering, Zhejiang University of Science and Technology, Liuhe Road 318, Hangzhou 310023, China
*   Correspondence: zhaogch@sdas.org (G.Z.); guohanjie@ustb.edu.cn (H.G.); yanpengxue@ustb.edu.cn (Y.X.)
†   These authors contributed equally to this work.

Received: 22 May 2019; Accepted: 16 July 2019; Published: 18 July 2019

Abstract: Corrosion—reactions occuring between engineering materials and their environment—can cause material failure and catastrophic accidents, which have a serious impact on economic development and social stability. Recently, super-hydrophobic coatings have received much attention due to their effectiveness in preventing engineering materials from further corrosion. In this paper, basic principles of wetting properties and corrosion protection mechanism of super-hydrophobic coatings are introduced firstly. Secondly, the fabrication methods by electrochemical surface engineering—including electrochemical anodization, micro-arc oxidation, electrochemical etching, and deposition—are presented. Finally, the stabilities and future directions of super-hydrophobic coatings are discussed in order to promote the movement of such coatings into real-world applications. The objective of this review is to bring a brief overview of the recent progress in the fabrication of super-hydrophobic coatings by electrochemical surface methods for corrosion protection of engineering materials.

Keywords: super-hydrophobic coating; corrosion protection; electrochemical surface engineering; anodization; micro-arc oxidation; etching; electrodeposition; stability

1. Introduction

Engineering materials such as steel, aluminum (Al), magnesium (Mg), titanium (Ti), and their alloys are conventional metals employed in industry due to their significant mechanical and physical performance. However, in real service, corrosion may cause partial or complete destruction due to the chemical/electrochemical reactions of engineering materials and their environment [1]. Serious corrosion will not only lead to material failure, but also cause catastrophic accidents which may have serious impacts on economic development and social stability [2]. In developed countries, the national cost of corrosion generally represents approximately 1–5% of the gross national product (GNP) [3].

Raising awareness of the importance of corrosion protection, various methods are applied on corrosion control engineering, such as development of novel corrosion resistant materials, corrosion inhibitors, coating and surface modifications, as well as electrochemical cathodic protection. Among these methods, coating is a simple and effective way for corrosion protection with low cost.

In the past several years, various coating techniques have been studied—like chemical vapor deposition (CVD) [4], plasma treatment [5–8], thermal spray [9], electrochemical methods [10–18], sol–gel [19,20], magnetron sputtering [21–23], etc. The above-mentioned methods mostly involve post-treatment or high temperatures in the fabrication process, which can damage the coating and/or substrate with relatively low melting points [24]. Therefore, among various techniques, electrochemical methods are the better choice for corrosion protection due to the merits of mild processing conditions, low cost, and applicability for large-scale production.

Super-hydrophobic surfaces, inspired by the lotus leaf [25], play a significant role in the fundamental research of functional materials [26,27]. In general, super-hydrophobic surface has water contact angle (WCA) larger than 150° and water sliding angle (WSA) below 10°, which is regarded as nearly perfect non-wetting [28]. Due to their special wetting properties, super-hydrophobic surfaces have been widely investigated in broad potential applications, such as anti-corrosion [29–34], anti-fouling [35,36], anti-icing [37,38], self-cleaning [27,39,40], oil–water separation [41,42], energy conversion [43], and catalysis [44]. Especially, the unique water repellency property of super-hydrophobic surfaces makes them practical to serve as an effective barrier to prevent the engineering materials underneath from further corrosion.

In this review, the whole layout was organized as follows: the basic principle of wetting properties on solid surface and corrosion protection mechanism of super-hydrophobic coatings are firstly described in Section 2. The fabrication of super-hydrophobic coatings by electrochemical surface engineering including electrochemical anodization, micro-arc oxidation, electrochemical etching, and deposition are presented in Section 3. The corresponding advantages and disadvantages of each method are also summarized. In Section 4, stabilities of super-hydrophobic coatings are discussed. Finally, perspectives on further development of artificial coatings are presented in Section 5.

2. Basic Principle of Wetting Properties and Corrosion Protection Mechanism of Super-Hydrophobic Coating

2.1. Basic Principle of Wetting Properties

Wetting property is one of the important parameters of the solid surface. When liquid droplets contact a solid surface, an appropriate and stable angle of contact is formed on the liquid–solid interface, which is defined as contact angle. Thomas Young described the trigonometric relations between the surface tensions and the contact angle in 1805 in his essay [45], as can be expressed as

$$\gamma_s - \gamma_{sl} = \gamma_l \cdot \cos\theta \tag{1}$$

where γ_s, γ_l, γ_{sl} are the tensions of the solid surface, liquid surface, solid–liquid interface, respectively, and θ is the contact angle, shown by Figure 1a.

However, wetting properties of realistic surfaces are more complex since such surfaces are generally rough and chemically heterogeneous. Wenzel explained the wetting phenomenon as a thermodynamic process from the view point of specific surface energy [46]. The net energy decrease determined wetting speed; net energy decrease was greater for the rougher surface, which suggested surface wettability should be strengthen. Figure 1b shows Wenzel wetting state. A surface ratio named "roughness factor" (r) was introduced as the actual surface area divided by geometric surface area in his work. Then actual contact angle θ_A could be calculated by

$$\cos\theta_A = r \cos\theta \tag{2}$$

From the definition of roughness factor, r is always no less than 1. Therefore, as $0° < \theta < 90°$, then $\theta_A \leq \theta$, similarly as $90° < \theta < 180°$, then $\theta_A \geq \theta$. Using another word, wettability—including hydrophilicity and hydrophobicity—can be enhanced by surface roughness increasement. When actual WCA is greater than 150°, the surface shows super-hydrophobicity.

Although Wenzel provided a strategy in fabricating superwetting surface, i.e., increasing roughness, the solid–liquid contact area was also increased resulting in a "sticky" surface in the Wenzel state [47]. Moreover, Equation (2) breaks down at relatively large r ($|rcos\theta|>1$), since in such case air is trapped within the surface rough structures. Therefore, Cassie and Baxter developed another model to describe wetting on porous surface [48], extended Wenzel's statement, namely Cassie–Baxter wetting state as shown by Figure 1c. The liquid droplet was suspended on such surface due to the contact with the solid surface and gas phase [49]. Then apparent contact angle should be described as

$$\cos\theta_A = f_{sl} \cos\theta_y + f_l \cos\theta_g \qquad (3)$$

where, f_{sl} and f_l are contacting area fractions of solid–liquid interface and liquid–gas interface in wetting part. The absolute values of both f_{sl} and f_l are less than 1, and the sum of f_{sl} and f_l is equal to 1. θ_y is Young's contact angle and θ_g is liquid–gas contact angle (always 180°). Thus, Equation (3) can be substituted as

$$\cos\theta_A = f_{sl}\left(\cos\theta_y + 1\right) - 1 \qquad (4)$$

Cassie–Baxter equation tells us that introducing air pockets or reducing the solid–liquid contact fraction by increasing the surface roughness can forcibly enlarge the apparent contact angle, even if the solid material is naturally hydrophilic.

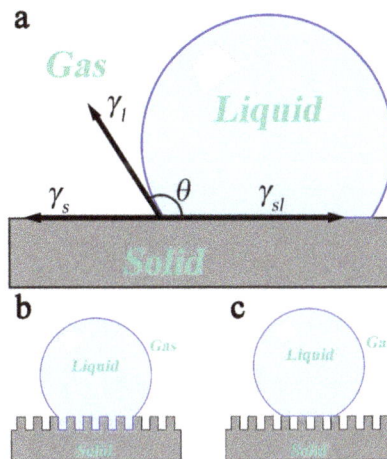

Figure 1. Schematic of a liquid droplet on solid surface. (**a**) Young's wetting state; (**b**) Wenzel wetting state; (**c**) Cassie–Baxter wetting state.

Super-hydrophobicity, introduced in 1976 by Reick [50], is a special wetting phenomenon which has extreme water-repellency. Normally, the super-hydrophobic surface has a WCA of at least 150°. However, contact angle can hardly exceed 120° on a smooth surface [51], even on extremely low surface energy materials, e.g., silicone resin of 22 mN/m, fluorine resin of 10 mN/m [52]. Barthlott and Neinhuis investigated the micro morphologies, surface chemicals of more than 200 water-repellent plant species and their self-cleaning abilities [25,53]. Demonstrated by the leaf of the *Nelumbo nucifera*, the waxy surface texture was indispensable in giving rise to water-repellency and self-cleaning of the plant. Water was suspended on the top of leaf rough structure and rolled off easily (evaluated by sliding angle) as well as remove any contaminants from the surface [28]. They named this mechanism the "lotus-effect", but the fundamental mechanism was not mentioned [25]. The super-hydrophobic mechanism was clearly explained in 2002, the 'lotus-effect' with high contact angle and low sliding angle, resulted from micro- and nanoscale hierarchical structures: nanostructures contributed to high contact angle and nano- and microstructures effectively reduced WSA [54].

2.2. Corrosion Protection Mechanism of Super-Hydrophobic Surface

Given their water repellency for corrosion protection, super-hydrophobic surfaces have been widely applied on many metals and alloys—e.g., aluminum, magnesium, steel, titanium, zinc, copper, and so on [49,55–60]. The corrosion resistance mechanism of super-hydrophobic surfaces can be summarized as follows: when super-hydrophobic surface immerses into a corrosive medium, an air layer is formed within the valleys among the rough structures between the super-hydrophobic surface and liquid phase, which minimized the contact area as a barrier remarkably hinders the corrosion of metal and alloy surface [61]. As illustrated by Figure 2, the air layer retained on super-hydrophobic surface prevents the chloride ions in seawater from attacking metal surface, therefore providing effect corrosion protection [62]. This mechanism has been widely demonstrated by electrochemical impedance spectroscopy (EIS). Super-hydrophobic surfaces always have high impedance modulus value in the Nyquist plots, representing high polarization resistance. The trapped air hindered the electron transfer between the metal and solution resulted in an effective corrosion protection for engineering materials.

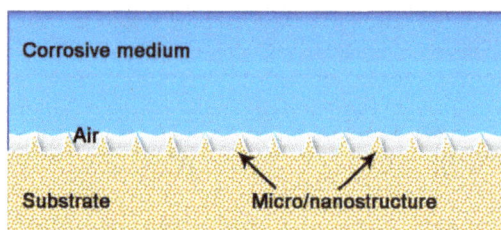

Figure 2. Schematic image of anti-corrosion using a super-hydrophobic surface. The super-hydrophobic micro/nanostructures keep surface wettability in 'Cassie State' to prevent corrosive medium penetrating the air layer and contacting with the substrate.

3. Electrochemical Surface Engineering

As it known to all that special micro-nanostructures and modification with low surface energy materials are pointed out to be the critical factors to fabricate the biomimetic super-hydrophobic coating. Among the applicable techniques, electrochemical methods offer outstanding advantages such as mild processing conditions, low-cost, time-efficiency, as well as large-scale production. Herein, several electrochemical methods—including electrochemical anodization, micro-arc oxidation, electrochemical etching, and electrochemical deposition—are introduced.

3.1. Electrochemical Anodization

Electrochemical anodization is a simple and effective approach that can be used to develop hierarchical micro/nanoscale oxide layer at the top surface of valve metals and these alloys [63]. The metal oxide layer is generated from the reaction between metal ions and the electrolytes under a high voltage provided by an external power source [64]. Their characteristics of micro/nanoscale oxide layer are strongly dependent on electrochemical parameters such as the electrolyte compositions (phosphoric acid, sulfuric acid, or oxalic acid) and temperature, applied voltage, anodization time, and distance between the working electrode and the counter electrode [65]. After producing hierarchical micro/nanoscale structures, modification with low surface energy materials is necessary to endow its super-hydrophobicity. Electrochemical anodization possesses many advantages for developing super-hydrophobic surfaces on valve metals because the production processes are easy to perform, highly efficient, relatively economical, and suitable for large-scale manufacturing.

Recently, numerous chemically-stable and mechanically-robust super-hydrophobic surfaces have been produced via the electrochemical anodization technique and subsequent modification process. Aluminum and titanium are the typical substrates for fabricating super-hydrophobic surfaces by this method. For example, Kondo et al. constructed super-hydrophobic surfaces on different kinds of aluminum alloys by electrochemical anodization in concentrated pyrophosphoric acid solution and subsequent chemical modification [66]. Electrochemical anodization of aluminum alloy could cause the growth of massive anodic aluminum nanofibers which were developed with the reaction time (Figure 3a). Figure 3b illustrates the surface morphology of pure aluminum anodized for different times. A labyrinthine structure with oxide walls was observed at the initial stage of anodization. When the anodizing time reached to 4 min and 10 min, honeycomb oxide structures and rod-shaped oxides were observed on the surface, respectively. Larger bundle structures were appealed on the surface after anodizing for 30 min. As the reaction time increased, the structures of the other two kinds of Al alloys gradually changed, similar to the transform of structures of pure aluminum. All Al alloys obtained the super-hydrophobicity after modified by low surface energy chemical reagents.

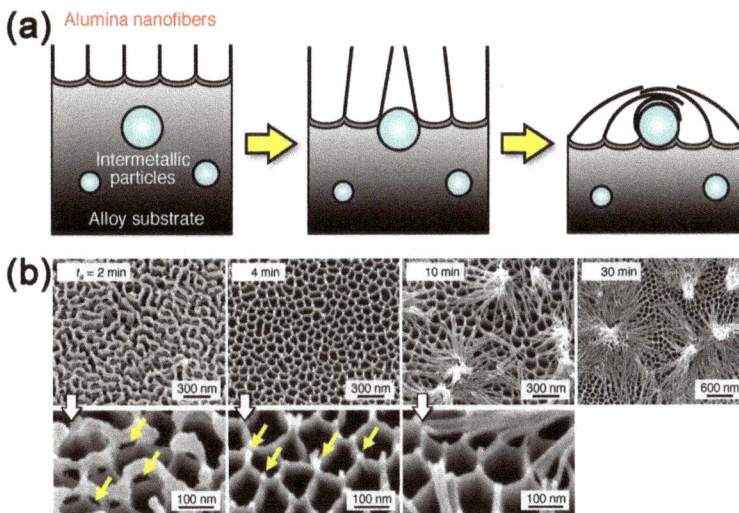

Figure 3. (**a**) Growth process of anodic aluminum nanofibers and nanofiber-tangled intermetallic particles; (**b**) SEM of samples anodized for t_a = 2 min, 4 min, 10 min, and 30 min, respectively. Reprinted from [66], Copyright (2017), with permission from Elsevier.

Zhang et al. fabricated a new anti-corrosion hierarchical alumina structure based on self-congregated nanowires through high-speed hard anodization technique on the high purity aluminum foils [29]. Figure 4a showed the fabricating process, the pure aluminum plates were pretreated first to obtain electropolished aluminum surface (EP surface) and then the anodization process was carried out in phosphoric acid solution under a constant voltage (120 V) to get super-hydrophilic (SHPL) surface. After that, the as-anodized sample was modified under the immersion of a 1% ethanol solution for 10 min and then dried at 120 °C. The as-prepared super-hydrophobic (SHPB) sample showed Dianthus caryophyllus-like structure (Figure 4b) and displayed WCA of 168°, showing excellent water repellency. The EIS results showed that the impedance modulus of SHPB surface was greater than 10^8 $\Omega \cdot cm^2$, which was much higher than that of EP and SHPL surface (Figure 4c). The corrosion inhibition efficiency was reached up to 99.99%, exhibiting the outstanding corrosion resistance of the SHPB surface. Besides, the as-received SHPB surface presented outstanding thermal stability to hot water droplets, which could retain contact angle higher than 150° under water droplets of 90 °C (Figure 4d).

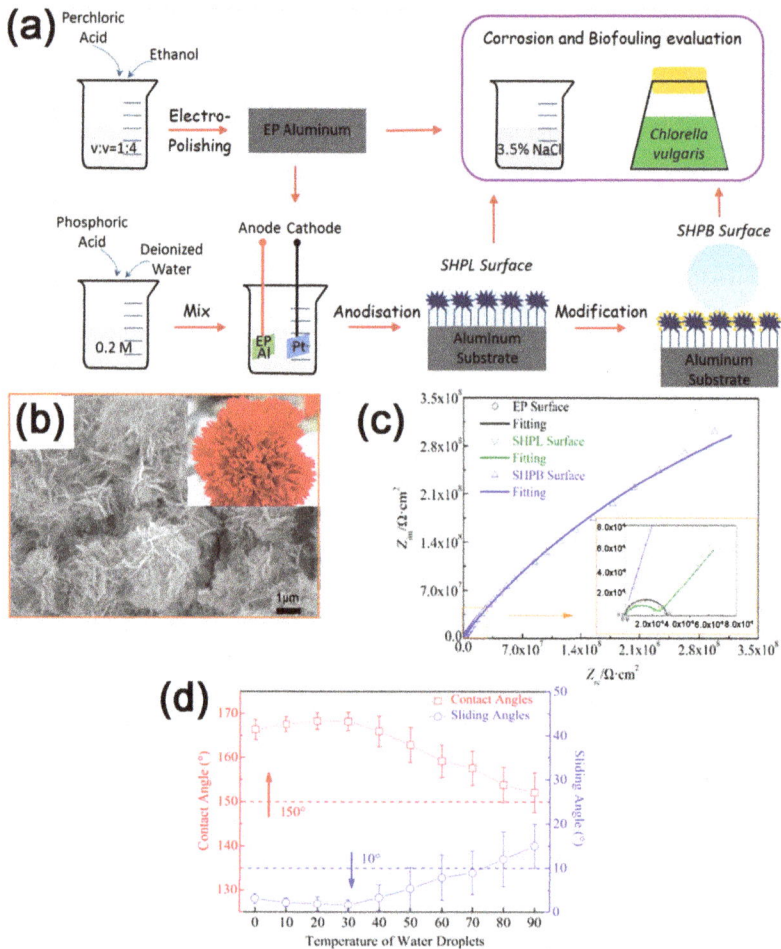

Figure 4. (a) Scheme of the preparation and evaluation of SHPL and SHPB surface by high-speed hard anodization; (b) SEM images of the *Dianthus caryophyllus*-like structure; (c) Nyquist plots and fittings of EP surface, SHPL surface, and SHPB surface; (d) The variations of WCAs and sliding angels of the SHPB surface under hot water droplets with different temperatures. Reprinted from [29], Copyright (2017), with permission from Elsevier.

Besides favorable corrosion resistance, abrasion performance is another important aspect needed to be considered in coating production. Peng and co-workers fabricated a robust super-hydrophobic aluminum surface possessing outstanding chemical stability and mechanical durability as well as resistance to many kinds of hot liquids. The coating was produced at 0.3 M oxalic acid electrolyte with a current density of 0.16 A/cm^2 within 10 min under room temperature, then modified with PDES (1H, 1H, 2H, 2H-perfluorodecyltriethoxysilane) [67]. After abrasion for 50 cycles, the nano-scaled structures of the PDES-modified surface were still retained and the WCA still remained above 150° both after ultra-sonication and abrasion test (Figure 5a–c). Moreover, the as-fabricated surface exhibited the improved corrosion resistance due to the promising water repellency. As shown by potentiodynamic polarization curves in Figure 5d, the corrosion current density (I_{corr}) of the PDES-modified surfaces, 1.31×10^{-11} A/cm^2, is two times of magnitude lower than that of the bare surfaces (9.70×10^{-8} A/cm^2).

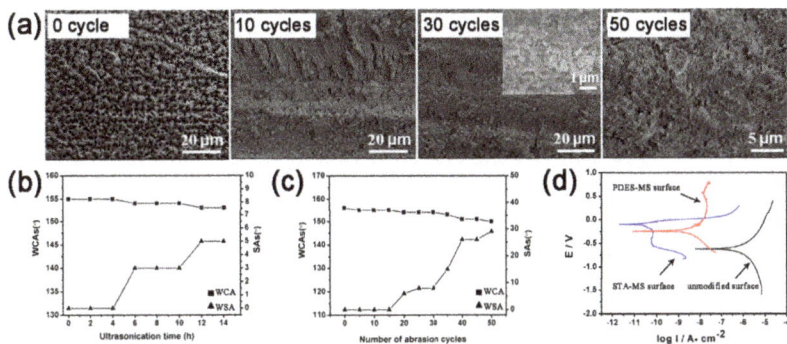

Figure 5. Mechanical stability and corrosion resistance evaluation of the PDES-MS surface: (**a**) Surface morphologies of the PDES-MS surface before and after abrasion test for 10, 30, and 50 cycles with sandpaper; (**b**) The variations of WCAs and WSAs under different ultrasonication time; (**c**) Variations of WCAs and WSAs of the PDES-MS surface after abrasion tests; (**d**) Potentiodynamic polarization curves of the unmodified and modified surfaces in 3.5 wt % NaCl solution. Reprinted with permission from [67]. Copyright (2014) American Chemical Society.

As is known, anodizing voltage has a great impact on the surface morphology of the coating. The effects of different anodizing voltages on the final contact angle have been investigated on a super-hydrophobic aluminum surface with prominent anticorrosive property via one-step anodization and chemically modification with stearic acid [68]. The WCA increased with the increase of the anodizing voltage until reached a peak at 25 V, and then declined. Excessive increase in voltage also leads to side effect on the size and order of the pores on the surface, which can decrease the uniform surface roughness. Combined with the suitable surface roughness of anodized aluminum and the following chemical modification, super-hydrophobic surface with contact angle of $152° ± 0.3°$ was received. At the optimized conditions, the result of electrochemical tests performed by 3.5 wt % NaCl solution showed that the I_{corr} of bare Al substrate and super-hydrophobic Al surface were 1.9 $\mu A/cm^2$ and 0.0562 $\mu A/cm^2$, respectively. The inhibition efficiency of super-hydrophobic Al surface was 97%, indicating the excellent corrosion resistance of super-hydrophobic surface.

Similar studies were conducted on anodized TiO_2 coatings [69]. Differ from the super-hydrophobic aluminum surface, the super-hydrophilicity of the Ti surfaces can be continuously enhanced with the increment of the anodization voltage from 20 V to 80 V without decline. Besides the anodizing voltage, parameters like electrolyte temperature also influence the final super-hydrophobicity. The increase in electrolyte temperature could improve the surface super-hydrophilicity in the low range of anodization voltages (<40 V). However, when the voltages rose up to 60 V, it showed the opposite trend. Suitable anodization voltage and electrolyte temperature can balance the growth and dissolution of the roughness surface structure, thereby producing the favorite superhydrophobic Ti surface. The WCA and WSA on the best super-hydrophobic TiO_2 surface were recorded as 160° and 2°, respectively. Moreover, the as-prepared super-hydrophobic Ti surfaces exhibited outstanding corrosion resistance in acidic, neutral, and alkaline aqueous solutions. The electrochemical reaction mechanism in super-hydrophobic Ti surface fabrication has been reported by Gao et al. [12]. They fabricated a super-hydrophobic Ti surface with a WCA of $158.5° ± 1.9°$ using anodization technique and subsequent modification with the fluoroalkylsilane (FAS). Anodization process was performed in a 1.5 mol/L NaOH and 0.15 mol/L H_2O_2 mixed solution at a DC voltage of 10 V within 30 min. During this process, the main electrochemical reaction was the anodic oxidation of Ti into TiO_2 and transformation of TiO_2 into Na_2TiO_3, as the equations

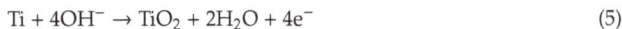

$$Ti + 4OH^- \rightarrow TiO_2 + 2H_2O + 4e^- \tag{5}$$

$$TiO_2 + 2NaOH \rightarrow Na_2TiO_3 + H_2O \tag{6}$$

The anodic oxidation reactions occurred immediately after the immersion of Ti alloy, and the chemical dissolution occurred due to the instability of TiO_2 in alkaline or acidic environment. The measurements by immersion and abrasion suggested that the as-received super-hydrophobic surfaces possessed good stability under various harsh conditions.

Combining electrochemical anodization with nano-silver deposition and post modification, Zhu et al. produced an anticorrosion super-hydrophobic film on Ti substrate [70]. With an appropriate immersion time of 7 h, micro-nano roughness structures was obtained. The WCA was reported to be 154°. Also, the corrosion resistance of super-hydrophobic surface was studied. As shown in Figure 6a, the specimen covered by the super-hydrophobic surface present much larger impedance semicircles whose diameter is around a few thousand of $k\Omega \cdot cm^2$. However, the pure Ti sample showed a small semicircle. At the frequency of 0.01 Hz, the super-hydrophobic sample displayed a high impedance modulus |Z| value of 1.319×10^3 $k\Omega \cdot cm^2$, which was nearly six times as large as that of the untreated Ti substrate (Figure 6b). Additionally, the super-hydrophobic sample presented more positive corrosion potential (E_{corr}) and lower I_{corr} compared with that of the untreated Ti (Figure 6c). The Nyquist plots and Bode plots, as well as polarization curves, demonstrated that the presence of the super-hydrophobic surface can effectively reduce the anodic dissolution of the Ti substrate and inhibit the development of corrosion.

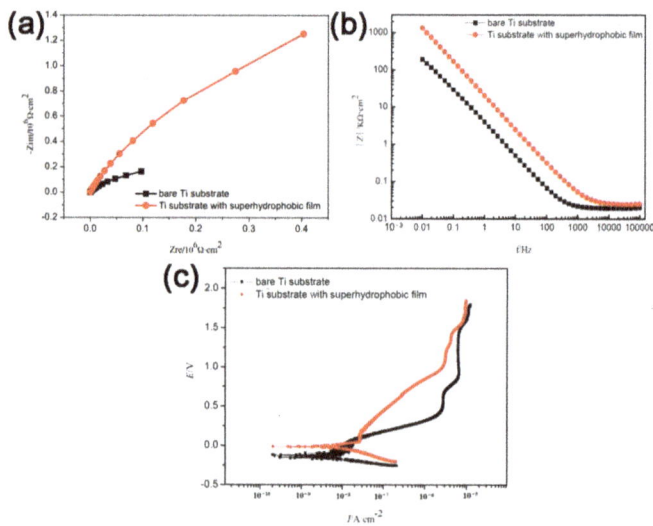

Figure 6. (a,b) Nyquist and Bode plots for the super-hydrophobic Ti surface and bare Ti substrate, respectively; (c) Polarization curves of super-hydrophobic Ti surface and bare Ti substrate. Reprinted from [70]. Copyright (2017), with permission from MDPI AG.

3.2. Micro-Arc Oxidation Coating

Micro-arc oxidation (MAO), also named as plasma electrolytic oxidation (PEO), is an electrochemical process to develop oxide coatings on metals. MAO employs high voltage plasma to modify the multi-layered ceramic structure which shows high hardness, high anti-corrosion behavior and strong adhesion to the metal surface [71–73]. The present of micro-pores on the coatings can provide the structure to fabricate super-hydrophobic surface. However, it can also increase the tendency to absorb the corrosive medium and penetrate into the substrate, which restrict its anti-corrosion applications [74]. Thus, in order to reduce or seal the micro-pores, many researchers combined MAO with other techniques.

Cui et al. fabricated a MAO/zinc stearate (ZnSA) composite coating with micro plate-like structure by MAO processing and electrochemical deposition method [75]. The hydrophilic coating was produced first by MAO in phosphate-containing electrolyte, then followed by the roughness control process and low surface energy material deposition to obtain the super-hydrophobic coating. After a short heating process, the MAO/ZnSA composite coating was received. As shown by Figure 7a,b, MAO coating with micro-pores provide mechanical connection for the super-hydrophobic coating and the MAO coating was sealed by the super-hydrophobic coating. The as-received white MAO coating shows a typical porous morphology with micro-pores and micro-cracks, while the surface of MAO/ZnSA coating exhibits whiter and rougher than the MAO sample (Figure 7b). The obtained super-hydrophobic coating showed a significant increase in contact angle of above 153° compare to the MAO of 37.5°. Based on the immersion tests, the MAO/ZnSA coated sample appealed no obvious corrosion pits after immersed in 3.5 wt % NaCl solution for 85 h, demonstrating excellent performance of corrosion resistance. Moreover, the relationship between the immersion time in NaCl and corresponding contact angles of the coating revealed that the contact angles change from 153.5° to 128.0° after immersion for 85 h, thus the long-term stability of the MAO coating need to be enhanced further.

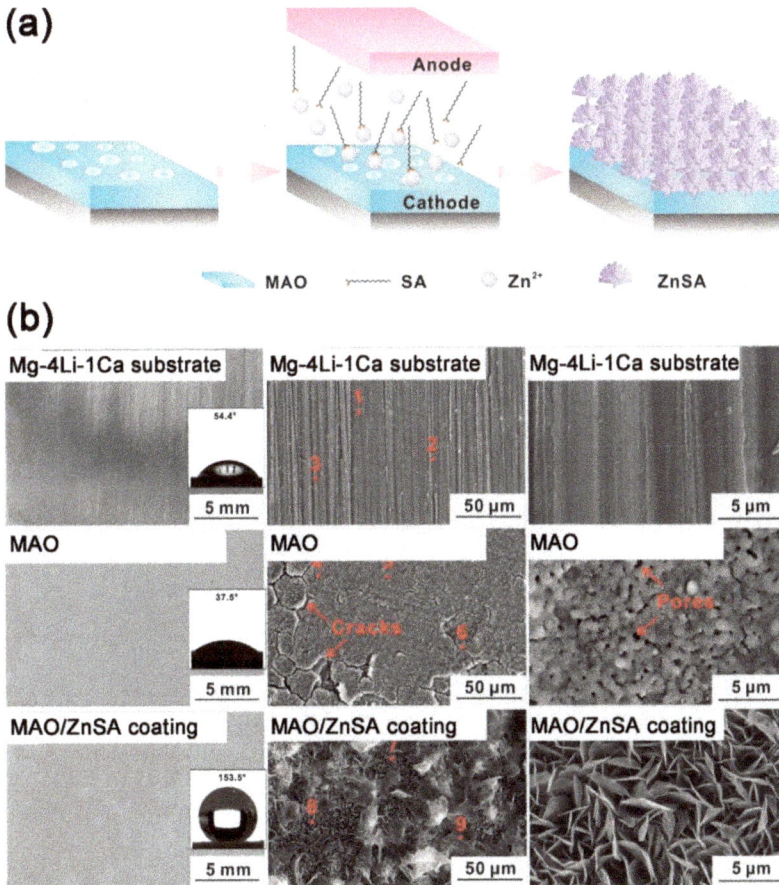

Figure 7. (**a**) Schematic of the electrodeposition process of MAO/ZnSA coating; (**b**) Macro-photographs and SEM images of Mg-4Li-1Ca substrate, MAO and MAO/ZnSA coating. The inserted are corresponding contact angles of each surface. Reprinted from [75]. Copyright (2017), with permission from Elsevier.

In another study, a super-hydrophobic surface on MAO coating on Mg-1Li-1Ca magnesium alloy with good performance of long-term stability was obtained by surface modification with stearic acid (SA) [76]. The porous structure of MAO coating was hydrophilic and the contact angle was below 35°. After SA modification, it was covered with thick petal-like clusters and exhibited super-hydrophobicity with contact angle over 155°, which was attributed to the high surface roughness of MAO coating as well as low surface energy of SA coating. As can be seen in Figure 8a, the sectional view of the MAO/SA coating showed a relatively dense and uniform structure with no obvious interface between the super-hydrophobic layer and the MAO coating, indicating that SA penetrating into the MAO layers and seals the micro-pores/cracks completely. Figure 8b presented the polarization curves of samples under different treatments, the I_{corr} of the substrate, MAO coating and MAO/SA coatings were in descending order and the I_{corr} of the MAO/SA coatings decreased with the extension of heating time. The I_{corr} of MAO/SA-7h coating was about 3 orders of magnitude lower compared with the bare substrate, suggesting a lower corrosion rate after surface modification. Also, the EIS results showed a significant increase in impedance modulus for MAO/SA coatings, revealing a higher corrosion resistance. Moreover, after 7 days' immersion in 3.5 wt % NaCl solution, the surface structures of the MAO/SA 7 h coating were still similar to the non-immersion specimen and the contact angle remained higher than 145°. As such, the fabricated MAO/SA coating demonstrated satisfactory stability and can perform well in long-term protection.

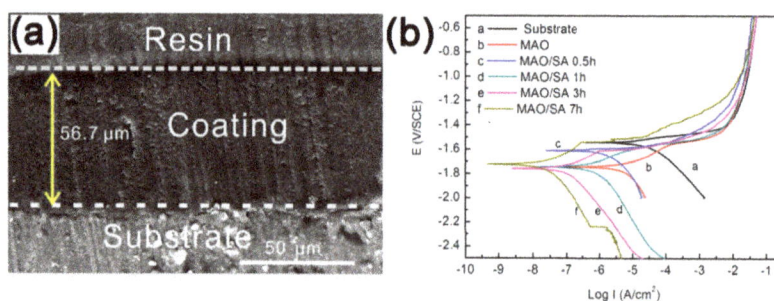

Figure 8. (**a**) Cross-view image of the MAO/SA 7h coating; (**b**) Polarization curves of the AZ31 substrate and the MAO coatings before and after modification with SA in 3.5 wt % NaCl solution. Reprinted from [76], Copyright (2017), with permission from Elsevier.

Similarly, Cui et al. [74] reported a simple process for synthesizing hydrophobic surface on AZ31 Mg alloy by MAO and surface modification of stearic acid. Combining with the rough micro-pore structures of MAO coating and the low surface energy of stearic acid monolayer, the super-hydrophobic surface with a maximum WCA over 151° was obtained after 10 h of modification (Figure 9a). The MAO-coated Mg alloy showed more positive E_{corr} and much lower I_{corr} than the uncoated substrate (Figure 9b), revealing a decreased corrosion susceptibility and a reduced corrosion rate.

The corrosion resistances of uncoated and coated alloys were investigated by EIS. As shown in Figure 9c, the MAO coating modified for 5 h (H-MAO) exhibited the largest diameter of the capacitive loop and no low-frequency inductive loop compared with the uncoated and unmodified coatings, which means a better corrosion resistance.

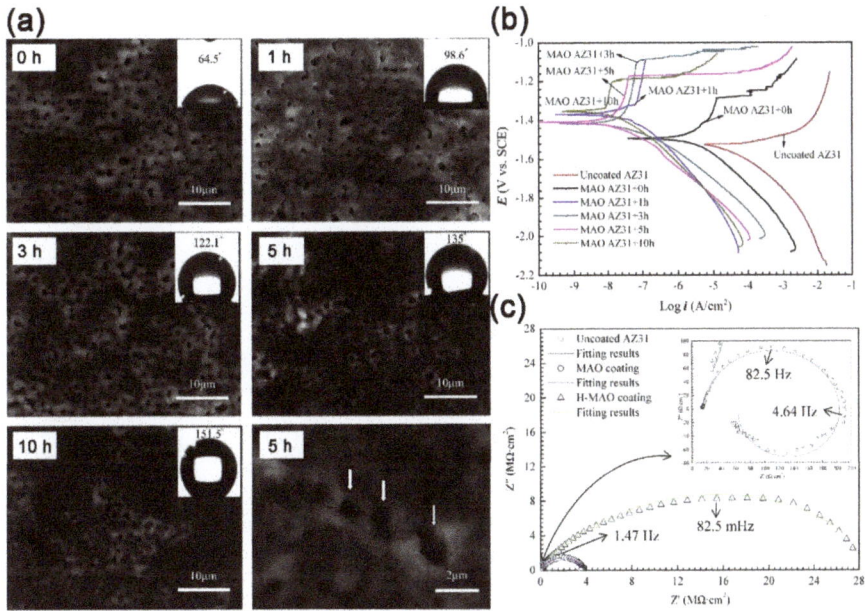

Figure 9. (a) Surface morphologies and inserted corresponding contact angles of the MAO coatings modified for 0, 1, 3, 5, and 10 h; (b) Potentiodynamic polarization curves of bare AZ31 Mg alloy and MAO coatings modified for different times; (c) EIS results and fitting curves for uncoated AZ31 Mg alloy, MAO coating, and H-MAO coating in 3.5 wt % NaCl solution. Reprinted from [74], Copyright (2015), with permission from Elsevier.

MAO method is also widely applied on Ti alloys. Jiang et al. [77] produced super-hydrophobic TiO$_2$ coatings on biomedical Ti-6Al-4V alloys by utilizing the MAO technique and super-hydrophobic treatment in 1H, 1H, 2H, 2H-perfluorooctyl-trichlorosilane (PFOTS) solution. The surface roughness Ra of the polished Ti-6Al-4V alloys was only 0.253 μm, while it increased remarkably to 0.535 μm after MAO treatment and further modification. Benefit from the increased surface roughness, the WCA of Ti-6Al-4V sample dropped slightly after MAO treatment and increased drastically after modification (Figure 10a), showing a high WCA over 153°. The potentiodynamic polarization curves shown by Figure 10b reveals that the E$_{corr}$ of super-hydrophobic sample was about 0.2 V nobler than the uncoated Ti-6Al-4V sample, while the I$_{corr}$ was reduced by one order of magnitude compared with that of uncoated sample, exhibiting an enhanced corrosion resistance. The super-hydrophobic TiO$_2$ coatings also exhibited better hemocompatibility and biocompatibility due to the reduced hemolysis ratio and platelets adhesion. As shown in Figure 10c, it is obvious that the surface of the uncoated alloy was adhered by a plenty of platelets while the platelets adhered on the surface of MAO sample reduced significantly and almost no platelet can be observed on the MAO + TFOS sample, indicating an enhanced hemocompatibility after MAO treatment and further super-hydrophobic treatment.

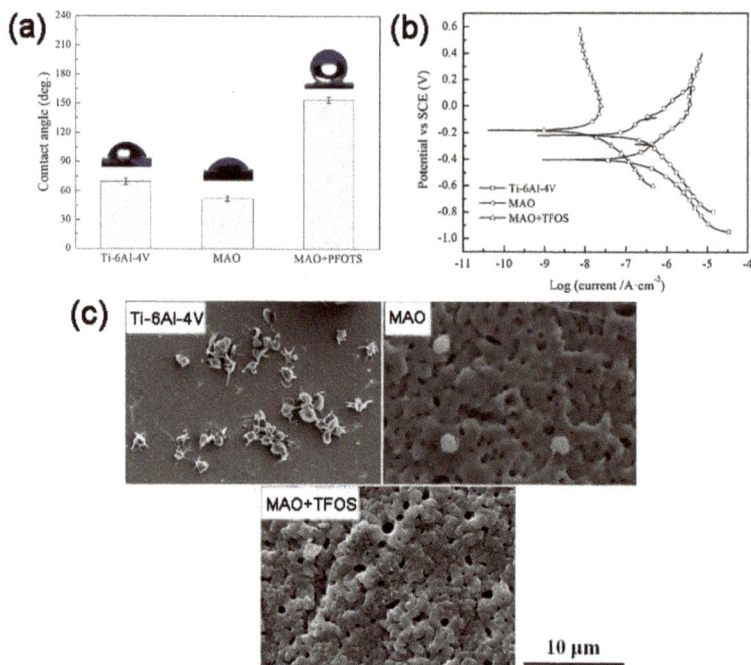

Figure 10. (**a,b**) Variations on contact angles and potentiodynamic polarization curves of the Ti-6Al-4V, MAO and MAO + PFOTS samples; (**c**) The morphologies of platelets adhered on the surface of the obtained samples. Reprinted from [77], Copyright (2015), with permission from Elsevier.

3.3. Electrochemical Etching

Etching is also a common technique to realize super-hydrophobic surfaces on various metals, such as titanium, stainless steel, aluminum and their alloys, etc. By electrochemical/chemical etching, hierarchical mirco-nano structures can be obtained. However, direct chemical etching is time consuming. Here, we mainly focus on electrochemical etching method.

For the preparation of super-hydrophobic titanium surfaces, Lu et al. reported the formation of surface microstructures via electrochemical etching in neutral sodium chloride electrolyte and subsequent modification by FAS to gain low surface energy [10]. The as-prepared Ti surfaces showed the super-hydrophobic properties with contact angles over 163°, and exhibited high stability and abrasion resistance. The same group also reported that the Ti surfaces can acquire both super-oleophobicity and super-hydrophobicity at the same time by adjusting the electrochemical etching parameters in NaBr solution [78]. Figure 11a–d showed that the droplets of the super-oleophobic and the super-hydrophobic Ti surface are both in spherical shape with contact angles greater than 150° in water and glycerol, while the liquids were spread out on unprocessed surface. This economical and environmentally-friendly electrochemical etching method is identified as a promising method suitable for industrial production of super-oleophobic and super-hydrophobic Ti surfaces for future applications in many fields.

Stainless steel (SS) is a basic alloy widely applied in industrial, living and medical fields due to its low price and reliable mechanical properties. With the growing need for high-performance SS, fabrication of super-hydrophobic coatings on SS with corrosion resistance properties arouse many researchers' concern. Recently, Song et al. developed a simple and low cost approach to prepare super-hydrophobic surfaces on mold steel GCr15 substrate by electrochemical etching and FAS modification with WCAs over 167° [79]. The bare substrates after electrochemical etching were covered with passive films composed of two-dimensional micro/nano rough structures, which is beneficial to fabricate super-hydrophobic surfaces. Jang et al. prepared a nanostructured stainless steel 316L (NT SS316L) surface by electrochemical etching for biomedical applications [80]. The NT SS316L surface showed nanopores with pore diameters between 20–25 nm which inhibited bacterial adhesion (Figure 12a,b). This electrochemical surface modification method can form outstanding passive layer on the surface of SS316L, which can improve corrosion resistance. The potentiodynamic polarization tests showed the as-received SS316Lsamples (AR-SS316L) exhibited localized corrosion with break down potential of 0.53 V (Figure 12c), while the local breakdowns of NT-SS316L samples were not observed. Additionally, the E_{corr} of NT SS316L is about 0.3 V nobler than the as-received SS316L sample, suggesting a less corrosive structure after surface modification.

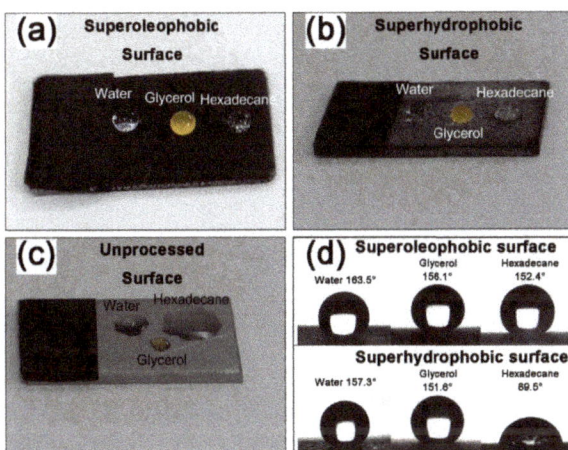

Figure 11. (a–c) Images of water droplets on the surface of superoleophobic Ti surface, super-hydrophobic Ti surface, and unprocessed Ti surface, respectively; (**d**) Variations in contact angles of super-oleophobic and super-hydrophobic Ti surface. Reprinted with permission from [78]. Copyright (2013) American Chemical Society.

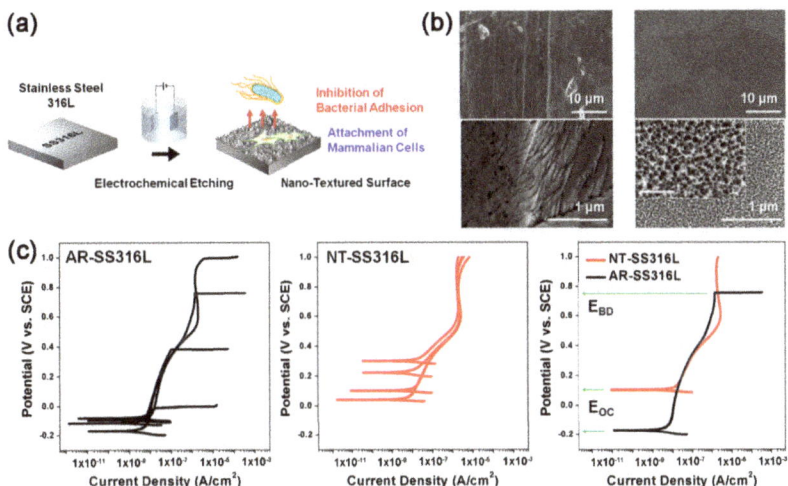

Figure 12. (**a**) Scheme of electrochemical etching method. (**b**) Surface morphologies of AR-SS316L and NT-SS316L surfaces. The scale bar of the inset image is 200 nm. (**c**) Potentiodynamic polarization curves of AR-SS316L and NT-SS316L specimens in Hank's balanced salt solution. Reprinted from [80]. Copyright (2017), with permission from American Chemical Society.

Other researchers also applied electrochemical etching method on aluminum surface. Yang et al. reported a two-step electrochemical-etching method combining electrochemical-etching and masking technology to fabricate super-hydrophilic dimple patterns on etched super-hydrophobic Al substrates [81]. The super-hydrophobic layer exhibited high E_{corr} due to the adhesion of the bubbles on the surface and the flat super-hydrophilic dimples can be controlled by switching different etching voltages. The findings will contribute to the design of new fog harvest devices. Chen et al. fabricated super-hydrophobic surfaces on aluminum film by electrochemical etching and myristic acid modification with WCA greater than 165° [82]. The results of potential polarization tests revealed that the anti-corrosion property of the super-hydrophobic surfaces were improved remarkably. Based on the data extracted from the polarization curves, after the super-hydrophobic treatment, the E_{corr} positively shifted from −0.87 V to −0.28 V and the I_{corr} dropped significantly for three times of magnitude, demonstrating that the Al coating had lower corrosion susceptibility and less corrosion rate in the natural environment.

3.4. Electrochemical Deposition

Currently, electrodeposition technique has aroused many concerns in fabricating super-hydrophobic surfaces due to its advantages compared with conventional coatings, such as simplicity, controllability, affordability, and ease of large-area modification. Moreover, the inherent connection of coating and base material is promising to enhance the mechanical robustness and corrosion properties, which is essential in practical applications.

For magnesium alloy, She et al. reported a robust and stable super-hydrophobic surface on AZ91D magnesium alloy by electrodeposition of nickel and chemical modification process [83]. The obtained surface showed pinecone-like hierarchical structure with WCA over 163° and can maintain the WCA above 150° after mechanical abrasion for 0.7 m under applied pressure of 1.2 kPa with 800 grit SiC sandpapers. It can still show super-hydrophobicity after long term expose in atmosphere for 240 days. The super-hydrophobic surfaces possess superior corrosion resistance in neutral 3.5 wt % NaCl solution, with only 0.003% of the corrosion rate of bare alloy in the potentiodynamic polarization test.

However, the above-mentioned two-step method requires the fabrication of a transition-metal layer and modification with the low-surface-energy material, which is complicated and time consuming. To solve this problem, Liu et al. developed a rapid one-step electrodeposition method to fabricate anticorrosion super-hydrophobic surface on Mg−Mn−Ce alloy by electrodepositing [11]. They studied the effect of deposition voltage and deposition time on the morphology and wettability of the coatings, finding that hierarchical structures were constructed at 30 V. The electrodeposition time can be as short as 1 min to obtain a super-hydrophobic surface and the papillae continued to agglomerate to form homogeneous hierarchical papillae structures until the electrodeposition time reached 20 min, all the contact angles were larger than 155° (Figure 13a). The super-hydrophobic surface exhibited strong corrosion resistance after immersion in corrosive aqueous solutions. The potentiodynamic polarization test results of the Mg alloy surfaces with and without treatment showed that all the E_{corr} shift nobly and I_{corr} decreased drastically after electrodeposition process (Figure 13b). Moreover, the Nyquist plots in Figure 13c confirmed that the super-hydrophobic surface showed much higher impedance values comparing with the bare alloy. To evaluate the mechanical stability, an abrasion test shown in Figure 13d was analyzed. The results revealed that the as-received surface can hold a contact angle over 150° after abrasion length of 0.4 m, suggesting a good mechanical durability.

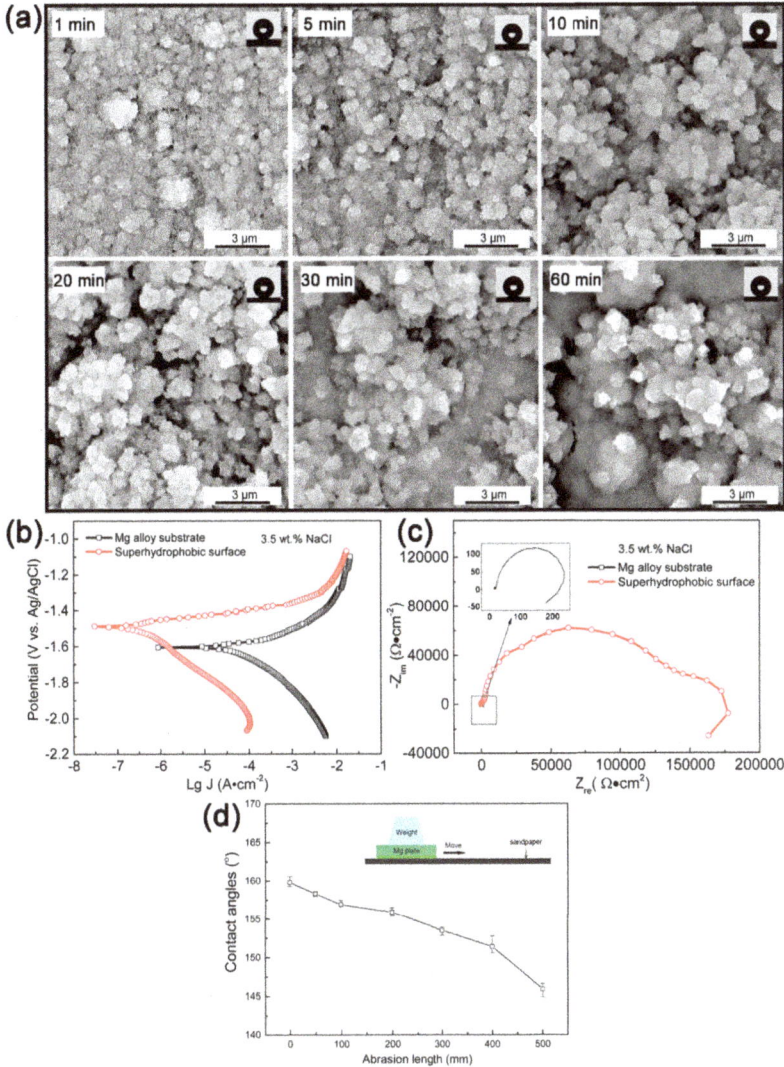

Figure 13. (**a**) SEM images and corresponding WCAs of the super-hydrophobic surfaces with various electrodeposition times: 1, 5, 10, 20, 30, and 60 min; (**b,c**) Potentiodynamic polarization curves and Nyquist plots of the untreated and super-hydrophobic Mg alloy surfaces in 3.5 wt % NaCl solution; (**d**) Schematic of the abrasion test and contact angles of the super-hydrophobic surface as a function of abrasion length. Reprinted with permission from [11]. Copyright (2015), American Chemical Society.

For controllable fabrication of super-hydrophobic surface on copper substrate, Su et al. [60] reported a novel and low-cost method by electrodeposition in traditional Watts bath and modification process (Figure 14a). The surface roughness increased from 0.08 μm for the Cu substrate to 1.86 μm for the deposited Ni surface, followed by a slightly decrease to 1.18 μm after surface modification process. The increased roughness of deposited layer contributed to the formation of superhydrophobic surface, which showed pine-cone-like hierarchical micro-nanostructure with a WCA of 162° ± 1°, possessed high microhardness and outstanding wear resistance after mechanical abrasion against 800 grit SiC sandpaper for 1.0 m at 2.4 kPa. When the applied pressure increased to 6.0 kPa, the pine-cone-like structure was partially damaged and the contact angle of the surface decreased (Figure 14b–e). Moreover, the super-hydrophobic surface exhibited good chemical stability both in acidic and alkaline environments. It is shown in Figure 14c that the Cu substrate showed higher E_{corr} and lower current density after Ni electrodeposition. The EIS test revealed that the impedance value of the super-hydrophobic surface is 75 times higher than that of the bare Cu substrate (Figure 14d), confirming the excellent corrosion protection properties of the super-hydrophobic surface.

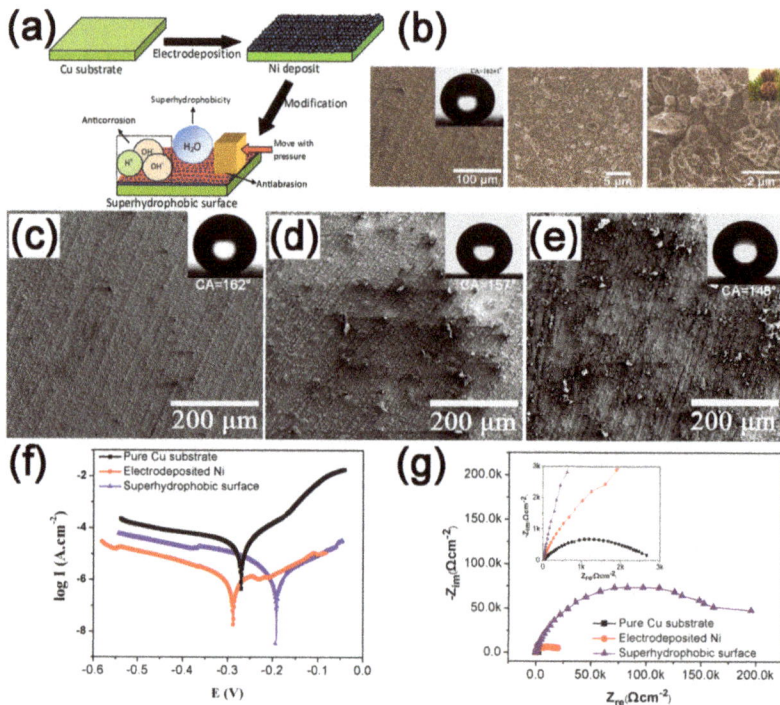

Figure 14. (a) Schematic of the electrochemical deposition process; (b) SEM images and contact angle of super-hydrophobic Ni (Ni-III) surface; (c–e) SEM images and contact angle of the super-hydrophobic surface (c) before and after abrasion for 1.0 m at applied pressure of (d) 2.4 kPa and (e) 6.0 kPa; (f,g) Potentiodynamic polarization curves and Nyquist plots of Cu substrate, electrodeposited Ni and the super-hydrophobic surface in 3.5 wt % NaCl solution. Reprinted with permission from [60]. Copyright (2014) American Chemical Society.

Similarly, a robust copper-based super-hydrophobic surface with cauliflower shaped fractal morphology was prepared by Jain et al. via an electrodeposition route [84]. The inherently generated super-hydrophobic surface showed extreme water repellency with CA above 160° (Figure 15a). The as-received coating exhibited slight reduction in E_{corr} and one order of magnitude lower I_{corr} in comparison to a bare copper substrate, indicating a significant increase of anticorrosion properties (Figure 15b). Moreover, the super-hydrophobic surface maintained integrity after the mechanical abrasion tests, showing favorable wear resistance (Figure 15c,d).

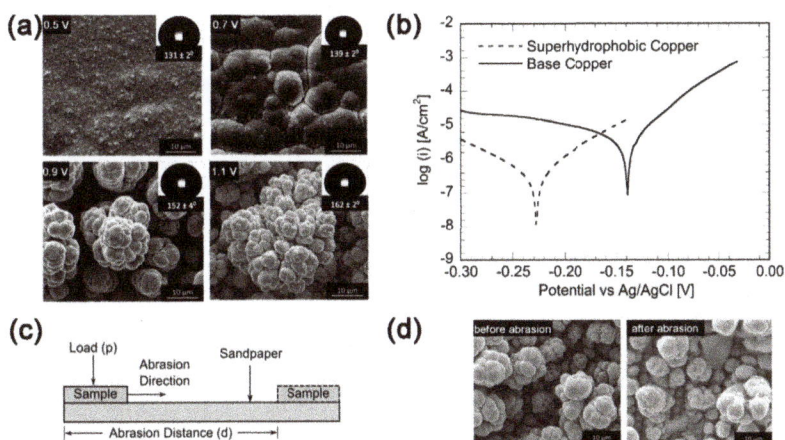

Figure 15. (**a**) SEM morphologies and WCAs of electrodeposited Cu fabricated at overpotentials of 0.5 V, 0.7 V, 0.9 V, and 1.1 V; (**b**) Potentiodynamic polarization curves for super-hydrophobic and base Cu in 3.5 wt % NaCl solution; (**c**) Schematic of the shear abrasion test setup; (**d**) SEM morphologies of super-hydrophobic sample before and after abrasion test. Reprinted (adapted) with permission from [84]. Copyright (2018) American Chemical Society.

4. Stability of Super-Hydrophobic Coating

So far, the fabricating strategies of super-hydrophobic surfaces have been developed a lot as discussed above. However, the instability of super-hydrophobicity seriously hinders its practical applications. Generally, super-hydrophobicity highly depends on both surface energy and micro-nano hierarchical structures. Such structures of surfaces are mechanically weak and stop functioning when facial chemicals are changed in various engineering conditions [85]. The surface energy could be easily increased with chemical attacks in aqueous acid, alkaline, salt solutions, or organic solvent. Meanwhile, the porous structures always get defects when mechanical damage applied on the surface. Additionally, normal ultraviolet (UV) irradiation and temperature variation may cause degradation of hydrophobic chemicals on surface. In this view, one of the future trends in developing super-hydrophobic surfaces with long-lasting corrosion resistance is to improve their stabilities [51]. Most recently published works have focused on improving the mechanical stability, chemical stability, and long-term durability of super-hydrophobicity. In this section, we discuss several types of stability as reference for the future fabrication of robust super-hydrophobic coatings.

4.1. Mechanical Stability

Improving mechanical stability is a key issue in fabricating robust super-hydrophobic surface. In other words, enhancing structural stability contributes to resist the external force to maintain original facial structures. The recently reported manners have been used to character such stability includes: linear abrasion [86], tape-peeling [55], knife scratching [87], finger pressing [88], bending [89], etc. Although so many approaches have been published for testing, there is still lack of standardization to make comparison of different surfaces. A general strategy to evaluate mechanical stability was suggested by Tian et al. (Figure 16). They reported that the linear abrasion test appears to best fulfill this requirement [90], which also is the most common applied manner to test robustness of super-hydrophobic surface in recent years [49,55,85–87,91–93].

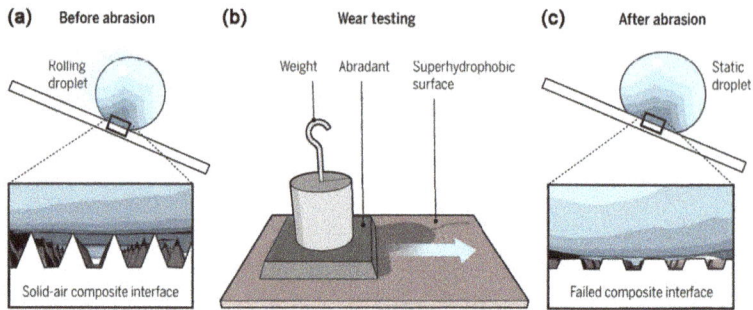

Figure 16. Linear abrasion test. (**a**) A water droplet rolls on a nonwetting surface; (**b**) The setup and process of linear abrasion test; (**c**) The structures are worn out after linear abrasion cause that the surface may lose water repellent property. Adapted with permission from. Reprinted from [90], Copyright (2016), with permission from AAAS.

For instance, Lu et al. created an ethanol-based suspension containing perfluorooctyltriethoxysilane coated dual-scale TiO_2 nanoparticles, which could be coated onto steel surface to create super-hydrophobicity by facile spray, dipping, or painting [85]. The paint directly coated on substrate could be easily removed by finger-wipe, whereas the double-sided tape-bonded paint (double-sided tape treated substrate and paint) could still retain its super-hydrophobicity after finger-wipe, knife-scratch, etc. Above all, the super-hydrophobicity of robust paint was not lost in the 0.4 m linear abrasion length (similar as Figure 16), which had potential in large-scale industrial applications. The linear abrasion test also applied on the surface of $Ni-WC-WS_2$ composite coating [94]. Such super-hydrophobic coating was fabricated by electrodeposition on mild steel with a WCA of approx. 170°. WS_2 nanoparticles (NPs) as solid self-lubricant could significantly reduce the coefficients of friction between solid–solid contacts, while WC NPs were hard materials to increase the abrasive resistance. Thus, the abrasive resistance of coating was further enhanced: with a bearing capacity ≥10,000 mm abrasion length on the 360# grit aluminum oxide paper under 3 kPa during the test.

Wang et al. carried out a facile oxidation to fabricated FAS-17 modified steel hierarchical surface [86]. The surface could withstand abrasion by 400 grids SiC sandpaper for 1.1 m under 16 kPa without losing super-hydrophobicity. Tam et al. applied co-electrodeposition process on nickel to synthesized a super-hydrophobic nanocrystalline nickel-polytetrafluoroethylene (Ni-PTFE) composite coating [95]. Due to the dual-scale surface roughness with lotus leaf-like morphology formed by the embedded PTFE particles in Ni matrix, the composite coatings possessed excellent wear resistance in abrasion test. The results in Figure 17a,b showed that, on 400 grit SiC sandpaper, the Ni-PTFE composites can only remain a high WCA above 150° for about 3 m of abrasion length and gradually decreased to about 130° after 18 m of abrasion. However, on the 800-grit sandpaper, stronger stability of super-hydrophobic properties was observed, the WCA of the Ni-PTFE composite coating can retain 150° after 48 m of abrasion under the applied pressure of 2.0 kPa. Xue et al. presented a electrodeposition route for super-hydrophobic Co–Ni coatings on carbon steel substrate which exhibited outstanding wear resistance [96]. The Co–Ni coating with micro-nano structures deposited at −1.7 V showed the highest surface roughness with Ra of 7.77 μm, which was much higher than that of −1.0 V (0.63 μm) and −1.4 V (1.71 μm). It shows that higher surface roughness structures can be favored to be generated by adjusting the applied potential to a more negative direction. The results of abrasion testing showed that the super-hydrophobic properties kept well after 12 m of abrasion under 5 kPa pressure (Figure 17c–f), which showed that the abrasion resistance was remarkably improved by the increase of Co content.

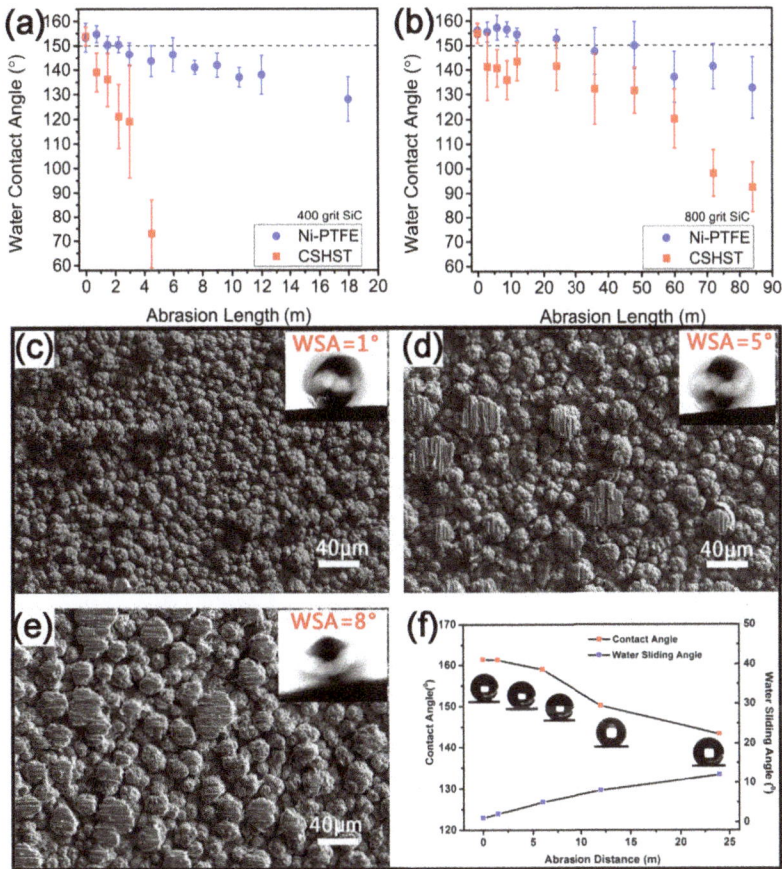

Figure 17. (**a,b**) Effect of abrasion length on WCA of Ni-PTFE composite coatings and CSHST coatings under 400 and 800 grit SiC papers, respectively. Reprinted from [95], Copyright (2017), with permission from Elsevier; (**c–e**) SEM images and inserted profiles of a water droplet for super-hydrophobic Co–Ni coating before abrasion and after abrasion wear of 6 m and 12 m under the applied pressure of 5 kPa, respectively; (**f**) Variations of the WCA and WSA on the tested surface with the abrasion distance [96]. Copyright (2019), with permission from MDPI AG.

Before being applied on a large-scale, super-hydrophobic surfaces should pass the examinations of other mechanical tests, such as hand twisting, tape-peeling, knife scratching. Test selection varies by the application of surface. Generally, particles or debris peeled off from the substrate indicates the structures are damaged and super-hydrophobicity may be failed [97]. Porous, loose, and uncompact surfaces can hardly bear such tear force (~300 N/m), whereas there is little influence on the surfaces like fluorination treated aluminum oxide [55], methyltrichlorosilane-Fe [89], FAS-17 modified rough steel [86], HVOF TiO$_2$/h-BN coating [97], and micro-nanostructure PDMS/SiO$_2$ composite coatings on Mg substrate [88], which demonstrates their robustness.

4.2. Chemical Stability

Most super-hydrophobic surfaces need to be modified by organic chemicals to minimize the surface energy. These surface modification agents might be degraded in harsh condition, which may cause increase of surface energy and decrease of super-hydrophobicity [87]. Thus, chemical stability of coating should be investigated in acid and base (or aqueous with various pH value), even aqua regia [87,98–101], salt aqueous [100–103], and organic solvent [85,87]. Besides, UV irradiation [86,104] and temperature variation [87,105] also should be taken into consideration which might cause degradation of surface chemicals.

To improve corrosion resistance of aluminum, Lv et al. used NaClO to the surface of aluminum and acquired super-hydrophobic surface after passivated by hexadecyltrimethoxysilane [103]. The sample maintained high contact angle (about 160.8°) after 7 days immersing in 3.5 wt % NaCl aqueous solution. The work reported by Qian et al. showed solution pH and salt aqueous including NaCl, MgCl$_2$, NaI, and CH$_3$COONa, have limited influence on the wetting properties of silica-based super-hydrophobic coating on AZ31B Mg Alloy [102]. Particularly, the organics like C$_2$H$_5$OH, C$_{17}$H$_{33}$COONa, and sodium dodecylbenzene sulfonate had obvious detrimental effects on the wetting properties. Khorsand et al. prepared super-hydrophobic nickel–cobalt alloy coating by a two-step-electrodeposition [100]. In the stability test, the coating was immersed into H$_2$SO$_4$ with pH of 2, NaOH with pH of 13 and neutral 3.5 wt % NaCl solution for 24 h. The results showed only in basic solution, the WCA decreased drastically with the increase of immersion time, which suggested the coating possessed favorable chemical stability in acidic and saline solution. Other products with super-hydrophobicity such as Zr-based MOFs [101] and silica gel [87] showed an excellent chemical stability in aqueous solution with various pH values ranging from 1 to 13, NaCl (0.5 M), Na$_2$SO$_4$ (0.5 M), n-hexane, xylene, butyl acetate, and acetone.

As for UV stability, UV exposure test was carried out for the super-hydrophobic steel surface [86]. The super-hydrophobic steel surface was prepared by FAS-17 modification of hierarchical steel. The C–F bonds of FAS-17 had a high bond energy (485 kJ/mol) which could not be broken by UV light. Thus, such surface still exhibited super-hydrophobicity with a contact angle of 151° and a sliding angle of 9° after 50 h of exposure under UV light, which suggested an excellent UV stability (Figure 18a). Xu et al. exploited a kind of transparent porous silica coating with the help of electrodeposited PEDOT template [106]. The coating showed super-hydrophobicity with excellent thermal stability. Its super-hydrophobicity remained constant in various temperatures ranging from 22 to 400 °C, see Figure 18b, and was lost when fluorosilane started decomposing at more than 400 °C.

Liu et al. reported super-hydrophobic fluorinated ZIF-90 showed high thermal stability in the temperature range 80–300 °C, and most bio-alcohol could be recovered and removed from mixture at 20 °C, which was very promising to be used as an effective and reusable adsorbent for bio-alcohols recovery from aqueous solution [105].

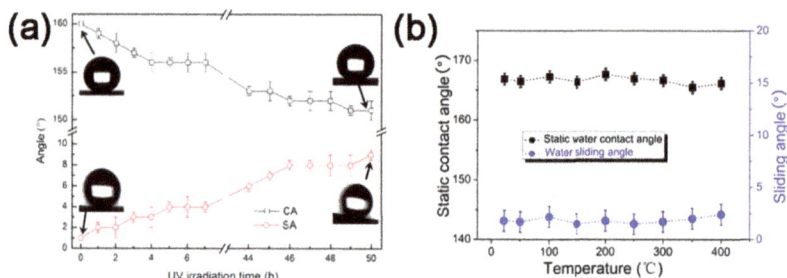

Figure 18. (**a**) WCAs and WSAs vary by different UV exposure time on super-hydrophobic steel. Reprinted (adapted) with permission from [86]. Copyright (2015) American Chemical Society; (**b**) The thermal stability of transparent porous silica coating. Reproduced with permission from [106]. Copyright (2015), with permission from Royal Society of Chemistry.

4.3. Long-Term Durability

The long-term durability is one of the most important properties for super-hydrophobic surface in actual applications, which determines the maximum lifespan of such functional surface. Generally speaking, the durability test is to measure the effective time of the surface having stayed in an unwetted state, which is always carried out at normal conditions, including atmospheric air [56,100,103,107], highly humid air [108], and water [107].

Ke et al. fabricated FAS-17 modified Al/Fe$_2$O$_3$ nanothermite film by electrophoretic deposition [108]. They put the film into humidity chamber with 60% relative humidity for 20 days. Due to infiltration of water vapor and air, the WCAs decreased slowly and the coating lost super-hydrophobicity after 7 days, but still showed strong hydrophobicity in the end of the test (about 145°). As for long-term water immersion, the copper-based super-hydrophobic blocks fabricated by Feng et al. [107] kept strong water repellent after immersing in water for a month, the WCAs of electrodeposited Ni-Co alloy super-hydrophobic coating on AA5052 aluminum increased from 151.3° to 160.0° during 15 weeks of exposure in open air, rather than decreased [109], and the HDYMS modified AA5083 aluminum could keep its water-repellent super-hydrophobicity stable in 12 days of 3.5 wt % NaCl immersion [110].

5. Perspectives

The mechanical durability and chemical stability of the as-prepared super-hydrophobic coatings needs to be further studied. Some possible directions to fabricate robust super-hydrophobic coatings are summarized as follows:

Self-healing coatings: One of the efficient ways to improve the durability and stability of the coatings is to endow it with self-healing ability. Generally, based on the healing mechanism, the repair process is divided into two categories: autonomous and non-autonomous. The healing effect and mechanism of the self-healing coatings have been elaborately discussed elsewhere [111]. For non-autonomous coatings, the external intervention, such as UV and heat, is indispensable to self-recovery. Under suitable stimuli, non-autonomous coatings can accurately repair the destroyed region. For the autonomous healing coatings, self-healing agents or inhibitors embedded in the coating layers can continuously provide corrosion protection at the defected region when the coating is damaged. For instance, corrosion inhibitors, along with polymerizable healing agents, can be mixed and packaged together in microcapsules. Figure 19a showed smart-coating-combined microcapsules containing linseed oil by the in situ polymerization method with polyelectrolytes layers entrapped benzotriazole (BTA) corrosion inhibitor. After application to the carbon steel surface, the coating exhibited self-healing property, which can release linseed oil under the stimulation of mechanical impact or release BTA by pH change [112]. Another novel study confirmed that the graphene oxides microcapsules containing linseed oil endowed self-healing properties to the waterborne polyurethane coatings (Figure 19b) [113]. Figure 19c,d depicted the self-healing process of the coating on cold-rolled steel, the rupture of microcapsules induced the polymerization of PDMS within the crack, resulting in recovery of corrosion protection [114].

Slippery liquid-infused porous surface (SLIPs): Inspired by Nepenthes pitcher plant, Wong et al. first designed slippery liquid-infused porous surface, which was conceptually different from the lotus effect [115]. In contrast with a super-hydrophobic surface with lotus-like structures, SLIPs use the micro/nanostructure to lock in place the infused lubricant instead of air (Figure 20). Therefore, SLIPs can withstand higher external pressure showing longer service life than the traditional super-hydrophobic surface. Additionally, SLIPs appealed outstanding liquid and ice-repellency, self-healing property, and corrosion resistance. These properties endow SLIPs with advantages in practical application for the corrosive environments.

Multi-functional: Water repellency and anti-corrosion is the already achieved function of the super-hydrophobic surfaces. The future super-hydrophobic surfaces must combine the above-mentioned properties with additional functions, such as super-oleophobicity and anti-biofouling. That is, the surface is multifunctional.

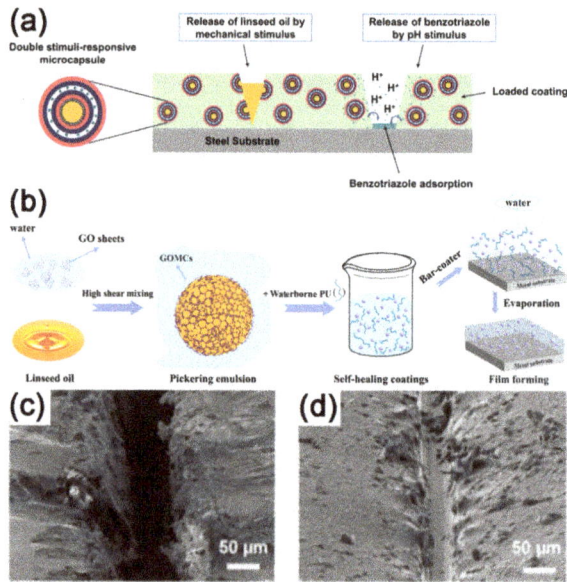

Figure 19. (**a**) Schematic of the structure of the microcapsules and the self-healing mechanisms of the coating. Reprinted from [112] Copyright (2018), with permission from Elsevier; (**b**) Formation process of oil-containing graphene oxide microcapsules (GOMCs) and the processing of GOMCs/PU coatings. Reprinted from [113], Copyright (2017), with permission from Elsevier; (**c**,**d**) SEM images of self-healing coating before and after healing. Reprinted from [114], Copyright (2009), with permission from John Wiley and Sons.

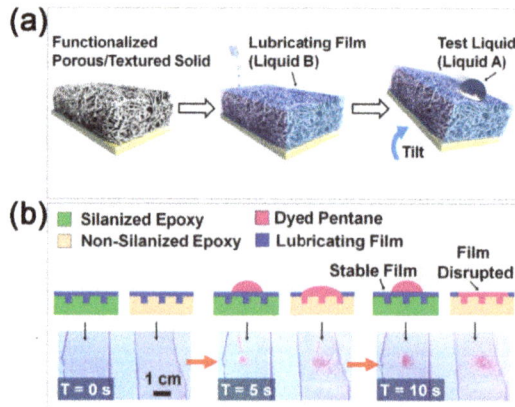

Figure 20. (**a**) Schematics of the fabrication process of a SLIPS; (**b**) Schematics and time-lapse images showing the stability and displacement of SLIPs. Reprinted from [115], Copyright (2011), with permission from Springer Nature.

6. Conclusions

Super-hydrophobic surfaces have received continuous attention in corrosion protection engineering, especially for long-time service safety under sensitive medium. Basic fundamentals for fabricating super-hydrophobic surfaces lie on the formation of rough micro/nanostructures and low surface energy materials. The corrosion protection of the super-hydrophobic surface is obtained by the air layer forms between the rough structures, which acts as a barrier between the alloy surface and corrosive medium.

In this review, we summarized the most currently used methods and techniques—including electrochemical anodization, micro-arc oxidation, electrochemical etching and deposition—to achieve surface roughness. After modification, the corrosion resistance and durability of super-hydrophobicity coatings were improved compared with bare substrate. Moreover, the mechanical and chemical stability of super-hydrophobic coating were discussed while the long-term stability of them still need to be enhanced. The development of robust super-hydrophobic coatings with anti-corrosion properties lies on the recent progress in self-healing coatings, slippery liquid-infused porous surface, as well as multi-functional coatings.

Funding: This research was funded by Fundamental Research Funds for the Central Universities China, grant number FRF-TP-18-009A1 and Shandong Province Key Research and Development Plan, grant number 2017CXGC0404.

Conflicts of Interest: The authors declare no conflict of interest.

References

1. Chang, R.M.; Kauffman, R.J.; Kwon, Y. Understanding the Paradigm Shift to Computational Social Science in the Presence of Big Data. *Decis. Support Syst.* **2014**, *63*, 67–80. [CrossRef]
2. Hou, B.; Lu, D. Corrosion Cost and Preventive Strategies in China. *Strategy Policy Decis. Res.* **2018**, *33*, 601–609. [CrossRef]
3. Hou, B.; Li, X.; Ma, X.; Du, C.; Zhang, D.; Zheng, M.; Xu, W.; Lu, D.; Ma, F. The Cost of Corrosion in China. *NPJ Mater. Degrad.* **2017**, *1*, 4. [CrossRef]
4. Pu, N.; Shi, G.; Liu, Y.; Sun, X.; Chang, J.; Sun, C.; Ger, M.; Chen, C.; Wang, P.; Peng, Y.; et al. Graphene Grown on Stainless Steel as a High-Performance and Ecofriendly Anti-Corrosion Coating for Polymer Electrolyte Membrane Fuel Cell Bipolar Plates. *J. Power Sources* **2015**, *282*, 248–256. [CrossRef]
5. Shao, F.; Zhao, H.; Zhong, X.; Zhuang, Y.; Cheng, Z.; Wang, L.; Tao, S. Characteristics of Thick Columnar YSZ Coatings Fabricated by Plasma Spray-Physical Vapor Deposition. *J. Eur. Ceram. Soc.* **2018**, *38*, 1930–1937. [CrossRef]
6. Song, C.; Liu, M.; Deng, Z.; Niu, S.; Deng, C.; Liao, H. A Novel Method for In-Situ Synthesized TiN Coatings by Plasma Spray-Physical Vapor Deposition. *Mater. Lett.* **2018**, *217*, 127–130. [CrossRef]
7. Zhang, B.; Song, W.; Guo, H. Wetting, Infiltration and Interaction Behavior of CMAS towards Columnar YSZ Coatings Deposited by Plasma Spray Physical Vapor. *J. Eur. Ceram. Soc.* **2018**, *38*, 3564–3572. [CrossRef]
8. Minařík, M.; Wrzecionko, E.; Minařík, A.; Grulich, O.; Smolka, P.; Musilová, L.; Junkar, I.; Primc, G.; Ptošková, B.; Mozetič, M.; et al. Preparation of Hierarchically Structured Polystyrene Surfaces with Superhydrophobic Properties by Plasma-Assisted Fluorination. *Coatings* **2019**, *9*, 201. [CrossRef]
9. Kear, B.H.; Kalman, Z.; Sadangi, R.K.; Skandan, G.; Colaizzi, J.; Mayo, W.E. Plasma-Sprayed Nanostructured Al₂/TiO₂. *J. Therm. Spray Technol.* **2000**, *4*, 483–487. [CrossRef]
10. Lu, Y.; Xu, W.; Song, J.; Liu, X.; Xing, Y.; Sun, J. Preparation of Superhydrophobic Titanium Surfaces via Electrochemical Etching and Fluorosilane Modification. *Appl. Surf. Sci.* **2012**, *263*, 297–301. [CrossRef]
11. Liu, Q.; Chen, D.; Kang, Z. One-Step Electrodeposition Process to Fabricate Corrosion-Resistant Superhydrophobic Surface on Magnesium Alloy. *ACS Appl. Mater. Interfaces* **2015**, *7*, 1859–1867. [CrossRef] [PubMed]
12. Gao, Y.; Sun, Y.; Guo, D. Facile Fabrication of Superhydrophobic Surfaces with Low Roughness on Ti-6Al-4V Substrates Via Anodization. *Appl. Surf. Sci.* **2014**, *314*, 754–759. [CrossRef]
13. Xue, Y.; Wang, S.; Zhao, G.; Taleb, A.; Jin, Y. Fabrication of Ni Co Coating by Electrochemical Deposition with High Super-Hydrophobic Properties for Corrosion Protection. *Surf. Coat. Technol.* **2019**, *363*, 352–361. [CrossRef]
14. Ban, C.; Shao, X.; Ma, J.; Chen, H. Effect of Mechanical Attrition on Microstructure and Property of Electroplated Ni Coating. *Trans. Nonferrous Met. Soc.* **2012**, *22*, 1989–1994. [CrossRef]
15. Ban, C.; Shao, X.; Wang, L. Ultrasonic Irradiation Assisted Electroless Ni-P Coating on Magnesium Alloy. *Surf. Eng.* **2014**, *30*, 880–885. [CrossRef]

16. Ban, C.; Wang, F.; Chen, J.; Zhu, S. Effect of Mechanical Attrition on Structure and Property of Electroplated Ni-P Coating on Magnesium Alloy. *Electrochemistry* **2019**, *87*, 89–93. [CrossRef]

17. Ban, C.; Hou, J.; Zhu, S.; Wang, C. Formation and Properties of Al_2O_3-ZrO_2 Composite Anodic Oxide Film on Etched Aluminum Foil by Electrodeposition and Anodization. *J. Mater. Sci. Mater. Electron.* **2016**, *27*, 1547–1552. [CrossRef]

18. Ban, C.; He, Y.; Shao, X.; Wang, L. Anodizing of Etched Aluminum Foil Coated with Modified Hydrous Oxide Film for Aluminum Electrolytic Capacitor. *J. Mater. Sci. Mater. Electron.* **2014**, *25*, 128–133. [CrossRef]

19. Chen, X.H.; Chen, C.S.; Xiao, H.N.; Cheng, F.Q.; Zhang, G.; Yi, G.J. Corrosion Behavior of Carbon nanotubes-Ni Composite Coating. *Surf. Coat. Technol.* **2005**, *191*, 351–356. [CrossRef]

20. Yang, X.F.; Tallman, D.E.; Gelling, V.J.; Bierwagen, G.P.; Kasten, L.S.; Berg, J. Use of a Sol-Gel Conversion Coating for Aluminum Corrosion Protection. *Surf. Coat. Technol.* **2001**, *140*, 44–50. [CrossRef]

21. Weng, J.; Zuo, X.; Liu, L.; Wang, Z.; Ke, P.; Wei, X.; Wang, A. Dense Nanocolumnar Structure Induced Anti-Corrosion CrB_2 Coating with (001) Preferred Orientation Deposited by DC Magnetron Sputtering. *Mater. Lett.* **2019**, *240*, 180–184. [CrossRef]

22. Thangavel, E.; Dhandapani, V.S.; Dharmalingam, K.; Marimuthu, M.; Veerapandian, M.; Arumugam, M.K.; Kim, S.; Kim, B.; Ramasundaram, S.; Kim, D. RF Magnetron Sputtering Mediated NiTi/Ag Coating on Ti-alloy Substrate with Enhanced Biocompatibility and Durability. *Mater. Sci. Eng. C* **2019**, *99*, 304–314. [CrossRef] [PubMed]

23. Yi, P.; Zhu, L.; Dong, C.; Xiao, K. Corrosion and Interfacial Contact Resistance of 316L Stainless Steel Coated with Magnetron Sputtered ZrN and TiN in the Simulated Cathodic Environment of a Proton-Exchange Membrane Fuel Cell. *Surf. Coat. Technol.* **2019**, *363*, 198–202. [CrossRef]

24. Feng, L.; Zhang, H.; Wang, Z.; Liu, Y. Superhydrophobic Aluminum Alloy Surface: Fabrication, Structure, and Corrosion Resistance. *Colloids Surf. A Physicochem. Eng. Asp.* **2014**, *441*, 319–325. [CrossRef]

25. Barthlott, W.; Neinhuis, C. Purity of the Sacred Lotus, or Escape from Contamination in Biological Surfaces. *Planta* **1997**, *202*, 1–8. [CrossRef]

26. Ma, M.; Hill, R.M. Superhydrophobic Surfaces. *Curr. Opin. Colloid Interface Sci.* **2006**, *11*, 193–202. [CrossRef]

27. Bhushan, B.; Jung, Y.C. Natural and Biomimetic Artificial Surfaces for Superhydrophobicity, Self-Cleaning, Low Adhesion, and Drag Reduction. *Prog. Mater. Sci.* **2011**, *56*, 1–108. [CrossRef]

28. Roach, P.; Shirtcliffe, N.J.; Newton, M.I. Progess in Superhydrophobic Surface Development. *Soft Matter* **2008**, *4*, 224–240. [CrossRef]

29. Zhang, B.; Hu, X.; Zhu, Q.; Wang, X.; Zhao, X.; Sun, C.; Li, Y.; Hou, B. Controllable Dianthus Caryophyllus-Like Superhydrophilic/Superhydrophobic Hierarchical Structure Based on Self-Congregated Nanowires for Corrosion Inhibition and Biofouling Mitigation. *Chem. Eng. J.* **2017**, *312*, 317–327. [CrossRef]

30. Cho, E.; Chang-Jian, C.; Chen, H.; Chuang, K.; Zheng, J.; Hsiao, Y.; Lee, K.; Huang, J. Robust Multifunctional Superhydrophobic Coatings with Enhanced Water/Oil Separation, Self-Cleaning, Anti-Corrosion, and Anti-Biological Adhesion. *Chem. Eng. J.* **2017**, *314*, 347–357. [CrossRef]

31. Liu, T.; Chen, S.; Cheng, S.; Tian, J.; Chang, X.; Yin, Y. Corrosion Behavior of Super-Hydrophobic Surface on Copper in Seawater. *Electrochim. Acta* **2007**, *52*, 8003–8007. [CrossRef]

32. Barkhudarov, P.M.; Shah, P.B.; Watkins, E.B.; Doshi, D.A.; Brinker, C.J.; Majewski, J. Corrosion Inhibition Using Superhydrophobic Films. *Corros. Sci.* **2008**, *50*, 897–902. [CrossRef]

33. Yu, S.Q.; Ling, Y.H.; Wang, R.G.; Zhang, J.; Qin, F.; Zhang, Z.J. Constructing Superhydrophobic WO_3@TiO_2 Nanoflake Surface Beyond Amorphous Alloy Against Electrochemical Corrosion on Iron Steel. *Appl. Surf. Sci.* **2018**, *436*, 527–535. [CrossRef]

34. Hsu, C.; Nazari, M.H.; Li, Q.; Shi, X. Enhancing Degradation and Corrosion Resistance of AZ31 Magnesium Alloy through Hydrophobic Coating. *Mater. Chem. Phys.* **2019**, *225*, 426–432. [CrossRef]

35. Oldani, V.; Del Negro, R.; Bianchi, C.L.; Suriano, R.; Turri, S.; Pirola, C.; Sacchi, B. Surface Properties and Anti-Fouling Assessment of Coatings Obtained from Perfluoropolyethers and Ceramic Oxides Nanopowders Deposited on Stainless Steel. *J. Fluor. Chem.* **2015**, *180*, 7–14. [CrossRef]

36. Ferrari, M.; Benedetti, A.; Cirisano, F. Superhydrophobic Coatings from Recyclable Materials for Protection in a Real Sea Environment. *Coatings* **2019**, *9*, 303. [CrossRef]

37. Farhadi, S.; Farzaneh, M.; Kulinich, S.A. Anti-Icing Performance of Superhydrophobic Surfaces. *Appl. Surf. Sci.* **2011**, *257*, 6264–6269. [CrossRef]

38. Cao, L.; Jones, A.K.; Sikka, V.K.; Wu, J.; Gao, D. Anti-Icing Superhydrophobic Coatings. *Langmuir* **2009**, *25*, 12444–12448. [CrossRef]

39. Fürstner, R.; Barthlott, W.; Neinhuis, C.; Walzel, P. Wetting and Self-Cleaning Properties of Artificial Superhydrophobic Surfaces. *Langmuir* **2005**, *21*, 956–961. [CrossRef]

40. Kadlečková, M.; Minařík, A.; Smolka, P.; Mráček, A.; Wrzecionko, E.; Novák, L.; Musilová, L.; Gajdošík, R. Preparation of Textured Surfaces on Aluminum-Alloy Substrates. *Materials* **2019**, *12*, 109. [CrossRef]

41. Song, J.; Huang, S.; Lu, Y.; Bu, X.; Mates, J.E.; Ghosh, A.; Ganguly, R.; Carmalt, C.J.; Parkin, I.P.; Xu, W.; et al. Self-Driven One-Step Oil Removal from Oil Spill on Water via Selective-Wettability Steel Mesh. *ACS Appl. Mater. Interfaces* **2014**, *6*, 19858–19865. [CrossRef]

42. Chen, N.; Pan, Q. Versatile Fabrication of Ultralight Magnetic Foams and Application for Oil-Water Separation. *ACS Nano* **2013**, *7*, 6875–6883. [CrossRef]

43. Yan, X.H.; Zhao, T.S.; An, L.; Zhao, G.; Zeng, L. A Crack-Free and Super-Hydrophobic Cathode Micro-Porous Layer for Direct Methanol Fuel Cells. *Appl. Energy* **2015**, *138*, 331–336. [CrossRef]

44. Shi, S.; Wang, M.; Chen, C.; Gao, J.; Ma, H.; Ma, J.; Xu, J. Super-Hydrophobic Yolk-Shell Nanostructure with Enhanced Catalytic Performance in the Reduction of Hydrophobic Nitroaromatic Compounds. *Chem. Commun.* **2013**, *49*, 9591. [CrossRef]

45. Young, T. An Essay on the Cohesion of Fluids. *Trans. R. Soc. Lond.* **1805**, 65–87. [CrossRef]

46. Wenzel, R.N. Resistance of Solid Surface to Wetting by Water. *Ind. Eng. Chem.* **1936**, *28*, 988–994. [CrossRef]

47. Tam, J.; Palumbo, G.; Erb, U. Recent Advances in Superhydrophobic Electrodeposits. *Materials* **2016**, *9*, 151. [CrossRef]

48. Cassie, A.B.D.; Baxter, S. Wettability of Porous Surfaces. *Trans. Faraday Soc.* **1944**, *40*, 546–551. [CrossRef]

49. Zhao, G.; Xue, Y.; Huang, Y.; Ye, Y.; Walsh, F.C.; Chen, J.; Wang, S. One-Step Electrodeposition of a Self-Cleaning and Corrosion Resistant Ni/WS$_2$ Superhydrophobic Surface. *RSC Adv.* **2016**, *6*, 59104–59112. [CrossRef]

50. Reick, F.G. Substrate Coated with Super-Hydrophobic Layers. U.S. Patent 3931428, 6 January 1976.

51. Zhang, D.; Wang, L.; Qian, H.; Li, X. Superhydrophobic Surfaces for Corrosion Protection: A Review of Recent Progresses and Future Directions. *J. Coat. Technol. Res.* **2016**, *13*, 11–29. [CrossRef]

52. Tuteja, A.; Choi, W.; Mabry, J.M.; McKinley, G.H.; Cohen, R.E. Robust Omniphobic Surfaces. *Proc. Natl. Acad. Sci. USA* **2008**, *105*, 18200–18205. [CrossRef]

53. Neinhuis, C.; Barthlott, W. Characterization and Distribution of Water-Repellent, Self-Cleaning Plant Surfaces. *Ann. Bot.-Lond.* **1997**, *79*, 667–677. [CrossRef]

54. Feng, L.; Li, S.; Li, Y.; Li, H.; Zhang, L.; Zhai, J.; Song, Y.; Liu, B.; Jiang, L.; Zhu, D. Super-Hydrophobic Surfaces: From Natural to Artificial. *Adv. Mater.* **2002**, *14*, 1857–1860. [CrossRef]

55. Lu, Z.; Wang, P.; Zhang, D. Super-Hydrophobic Film Fabricated on Aluminium Surface as a Barrier to Atmospheric Corrosion in a Marine Environment. *Corros. Sci.* **2015**, *91*, 287–296. [CrossRef]

56. Xu, W.; Song, J.; Sun, J.; Lu, Y.; Yu, Z. Rapid Fabrication of Large-Area, Corrosion-Resistant Superhydrophobic Mg Alloy Surfaces. *ACS Appl. Mater. Interfaces* **2011**, *3*, 4404–4414. [CrossRef]

57. Zhang, F.; Zhao, L.; Chen, H.; Xu, S.; Evans, D.G.; Duan, X. Corrosion Resistance of Superhydrophobic Layered Double Hydroxide Films on Aluminum. *Angew. Chem. Int. Ed.* **2008**, *47*, 2466–2469. [CrossRef]

58. Liu, H.; Szunerits, S.; Xu, W.; Boukherroub, R. Preparation of Superhydrophobic Coatings on Zinc as Effective Corrosion Barriers. *ACS Appl. Mater. Interfaces* **2009**, *1*, 1150–1153. [CrossRef]

59. Liu, T.; Yin, Y.; Chen, S.; Chang, X.; Cheng, S. Super-Hydrophobic Surfaces Improve Corrosion Resistance of Copper in Seawater. *Electrochim. Acta* **2007**, *52*, 3709–3713. [CrossRef]

60. Su, F.; Yao, K. Facile Fabrication of Superhydrophobic Surface with Excellent Mechanical Abrasion and Corrosion Resistance on Copper Substrate by a Novel Method. *ACS Appl. Mater. Interfaces* **2014**, *6*, 8762–8770. [CrossRef]

61. Wu, L.; Zhang, X.; Hu, J. Corrosion Protection of Mild Steel by One-Step Electrodeposition of Superhydrophobic Silica Film. *Corros. Sci.* **2014**, *85*, 482–487. [CrossRef]

62. Mohamed, A.M.A.; Abdullah, A.M.; Younan, N.A. Corrosion Behavior of Superhydrophobic Surfaces: A Review. *Arab J. Chem* **2015**, *8*, 749–765. [CrossRef]

63. Gao, A.; Hang, R.; Bai, L.; Tang, B.; Chu, P.K. Electrochemical Surface Engineering of Titanium-Based Alloys for Biomedical Application. *Electrochim. Acta* **2018**, *271*, 699–718. [CrossRef]

64. Vazirinasab, E.; Jafari, R.; Momen, G. Application of Superhydrophobic Coatings as a Corrosion Barrier: A Review. *Surf. Coat. Technol.* **2018**, *341*, 40–56. [CrossRef]

65. Darmanin, T.; de Givenchy, E.T.; Amigoni, S.; Guittard, F. Superhydrophobic Surfaces by Electrochemical Processes. *Adv. Mater.* **2013**, *25*, 1378–1394. [CrossRef]

66. Kondo, R.; Nakajima, D.; Kikuchi, T.; Natsui, S.; Suzuki, R.O. Superhydrophilic and Superhydrophobic Aluminum Alloys Fabricated Via Pyrophosphoric Acid Anodizing and Fluorinated SAM Modification. *J. Alloys Compd.* **2017**, *725*, 379–387. [CrossRef]

67. Peng, S.; Tian, D.; Yang, X.; Deng, W. Highly Efficient and Large-Scale Fabrication of Superhydrophobic Alumina Surface with Strong Stability Based on Self-Congregated Alumina Nanowires. *ACS Appl. Mater. Interfaces* **2014**, *6*, 4831–4841. [CrossRef]

68. Mokhtari, S.; Karimzadeh, F.; Abbasi, M.H.; Raeissi, K. Development of Super-Hydrophobic Surface on Al 6061 by Anodizing and the Evaluation of its Corrosion Behavior. *Surf. Coat. Technol.* **2017**, *324*, 99–105. [CrossRef]

69. Liang, J.; Liu, K.; Wang, D.; Li, H.; Li, P.; Li, S.; Su, S.; Xu, S.; Luo, Y. Facile Fabrication of Superhydrophilic/Superhydrophobic Surface On Titanium Substrate by Single-Step Anodization and Fluorination. *Appl. Surf. Sci.* **2015**, *338*, 126–136. [CrossRef]

70. Zhu, M.; Tang, W.; Huang, L.; Zhang, D.; Du, C.; Yu, G.; Chen, M.; Chowwanonthapunya, T. Preparation of Superhydrophobic Film on Ti Substrate and its Anticorrosion Property. *Materials* **2017**, *10*, 628. [CrossRef]

71. Yerokhin, A.L.; Nie, X.; Leyland, A.; Matthews, A.; Dowey, S.J. Plasma Electrolysis for Surface Engineering. *Surf. Coat. Technol.* **1999**, *122*, 73–93. [CrossRef]

72. Cai, J.; Cao, F.; Chang, L.; Zheng, J.; Zhang, J.; Cao, C. The Preparation and Corrosion Behaviors of MAO Coating on AZ91D with Rare Earth Conversion Precursor Film. *Appl. Surf. Sci.* **2011**, *257*, 3804–3811. [CrossRef]

73. Lin, C.S.; Fang, S.K. Formation of Cerium Conversion Coatings on AZ31 Magnesium Alloys. *J. Electrochem. Soc.* **2005**, *152*, B54. [CrossRef]

74. Cui, X.; Lin, X.; Liu, C.; Yang, R.; Zheng, X.; Gong, M. Fabrication and Corrosion Resistance of a Hydrophobic Micro-Arc Oxidation Coating on AZ31 Mg Alloy. *Corros. Sci.* **2015**, *90*, 402–412. [CrossRef]

75. Cui, L.; Liu, H.; Zhang, W.; Han, Z.; Deng, M.; Zeng, R.; Li, S.; Wang, Z. Corrosion Resistance of a Superhydrophobic Micro-Arc Oxidation Coating on Mg-4Li-1Ca Alloy. *J. Mater. Sci. Technol.* **2017**, *33*, 1263–1271. [CrossRef]

76. Zhang, C.L.; Zhang, F.; Song, L.; Zeng, R.C.; Li, S.Q.; Han, E.H. Corrosion Resistance of a Superhydrophobic Surface on Micro-Arc Oxidation Coated Mg-Li-Ca Alloy. *J. Alloys Compd.* **2017**, *728*, 815–826. [CrossRef]

77. Jiang, J.Y.; Xu, J.L.; Liu, Z.H.; Deng, L.; Sun, B.; Liu, S.D.; Wang, L.; Liu, H.Y. Preparation, Corrosion Resistance and Hemocompatibility of the Superhydrophobic TiO_2 Coatings on Biomedical Ti-6Al-4V Alloys. *Appl. Surf. Sci.* **2015**, *347*, 591–595. [CrossRef]

78. Lu, Y.; Song, J.; Liu, X.; Xu, W.; Xing, Y.; Wei, Z. Preparation of Superoleophobic and Superhydrophobic Titanium Surfaces via an Environmentally Friendly Electrochemical Etching Method. *ACS Sustain. Chem. Eng.* **2012**, *1*, 102–109. [CrossRef]

79. Song, J.; Huang, W.; Liu, J.; Huang, L.; Lu, Y. Electrochemical Machining of Superhydrophobic Surfaces on Mold Steel Substrates. *Surf. Coat. Technol.* **2018**, *344*, 499–506. [CrossRef]

80. Jang, Y.; Choi, W.T.; Johnson, C.T.; García, A.J.; Singh, P.M.; Breedveld, V.; Hess, D.W.; Champion, J.A. Inhibition of Bacterial Adhesion on Nanotextured Stainless Steel 316L by Electrochemical Etching. *ACS Biomater. Sci. Eng.* **2017**, *4*, 90–97. [CrossRef]

81. Yang, X.; Song, J.; Liu, J.; Liu, X.; Jin, Z. A Twice Electrochemical-Etching Method to Fabricate Superhydrophobic-Superhydrophilic Patterns for Biomimetic Fog Harvest. *Sci. Rep.* **2017**, *7*, 8816. [CrossRef]

82. Chen, Z.; Shuai, M.; Wang, L. Cathodic Etching for Fabrication of Super-Hydrophobic Aluminum Coating with Micro/Nano-Hierarchical Structure. *J. Solid State Electr.* **2013**, *17*, 2661–2669. [CrossRef]

83. She, Z.; Li, Q.; Wang, Z.; Li, L.; Chen, F.; Zhou, J. Researching the Fabrication of Anticorrosion Superhydrophobic Surface On Magnesium Alloy and its Mechanical Stability and Durability. *Chem. Eng. J.* **2013**, *228*, 415–424. [CrossRef]

84. Jain, R.; Pitchumani, R. Facile Fabrication of Durable Copper-Based Superhydrophobic Surfaces via Electrodeposition. *Langmuir* **2018**, *34*, 3159–3169. [CrossRef]

85. Lu, Y.; Sathasivam, S.; Song, J.; Crick, C.R.; Carmalt, C.J.; Parkin, I.P. Robust Self-Cleaning Surfaces that Function When Exposed to Either Air or Oil. *Science* **2015**, *347*, 1132–1135. [CrossRef]

86. Wang, N.; Xiong, D.; Deng, Y.; Shi, Y.; Wang, K. Mechanically Robust Superhydrophobic Steel Surface with Anti-Icing, UV-Durability, and Corrosion Resistance Properties. *ACS Appl. Mater. Interfaces* **2015**, *7*, 6260–6272. [CrossRef]

87. Ye, H.; Zhu, L.; Li, W.; Liu, H.; Chen, H. Simple Spray Deposition of a Water-Based Superhydrophobic Coating with High Stability for Flexible Applications. *J. Mater. Chem. A* **2017**, *5*, 9882–9890. [CrossRef]

88. Xie, J.; Hu, J.; Lin, X.; Fang, L.; Wu, F.; Liao, X.; Luo, H.; Shi, L. Robust and Anti-Corrosive PDMS/SiO₂ Superhydrophobic Coatings Fabricated on Magnesium Alloys with Different-Sized SiO₂ Nanoparticles. *Appl. Surf. Sci.* **2018**, *457*, 870–880. [CrossRef]

89. Latthe, S.S.; Sudhagar, P.; Devadoss, A.; Kumar, A.M.; Liu, S.; Terashima, C.; Nakata, K.; Fujishima, A. A Mechanically Bendable Superhydrophobic Steel Surface with Self-Cleaning and Corrosion-Resistant Properties. *J. Mater. Chem. A* **2015**, *3*, 14263–14271. [CrossRef]

90. Tian, X.; Verho, T.; Ras, R.H.A. Moving Superhydrophobic Surfaces Toward Real-World Applications. *Science* **2016**, 142–143. [CrossRef]

91. Gao, X.; Guo, Z. Mechanical Stability, Corrosion Resistance of Superhydrophobic Steel and Repairable Durability of its Slippery Surface. *J. Colloid Interface Sci.* **2018**, *512*, 239–248. [CrossRef]

92. He, Y.; Sun, W.T.; Wang, S.C.; Reed, P.A.S.; Walsh, F.C. An Electrodeposited Ni-P-WS₂ Coating with Combined Super-Hydrophobicity and Self-Lubricating Properties. *Electrochim. Acta* **2017**, *245*, 872–882. [CrossRef]

93. Zhao, G.; Li, J.; Huang, Y.; Yang, L.; Ye, Y.; Walsh, F.C.; Chen, J.; Wang, S. Robust Ni/WC Superhydrophobic Surfaces by Electrodeposition. *RSC Adv.* **2017**, *7*, 44896–44903. [CrossRef]

94. Zhou, J.; Zhao, G.; Li, J.; Chen, J.; Zhang, S.; Wang, J.; Walsh, F.C.; Wang, S.; Xue, Y. Electroplating of Non-Fluorinated Superhydrophobic Ni/WC/WS₂ Composite Coatings with High Abrasive Resistance. *Appl. Surf. Sci.* **2019**, *487*, 1329–1340. [CrossRef]

95. Tam, J.; Jiao, Z.; Lau, J.C.F.; Erb, U. Wear Stability of Superhydrophobic Nano Ni-PTFE Electrodeposits. *Wear* **2017**, *374–375*, 1–4. [CrossRef]

96. Xue, Y.; Wang, S.; Bi, P.; Zhao, G.; Jin, Y. Super-Hydrophobic Co-Ni Coating with High Abrasion Resistance Prepared by Electrodeposition. *Coatings* **2019**, *9*, 232. [CrossRef]

97. Zhang, F.; Robinson, B.W.; de Villiers-Lovelock, H.; Wood, R.J.K.; Wang, S.C. Wettability of Hierarchically-Textured Ceramic Coatings Produced by Suspension HVOF Spraying. *J. Mater. Chem. A* **2015**, *3*, 13864–13873. [CrossRef]

98. Zhang, X.; Zhi, D.; Sun, L.; Zhao, Y.; Tiwari, M.K.; Carmalt, C.J.; Parkin, I.P.; Lu, Y. Super-Durable, Non-Fluorinated Superhydrophobic Free-Standing Items. *J. Mater. Chem. A* **2018**, *6*, 357–362. [CrossRef]

99. Davis, A.; Surdo, S.; Caputo, G.; Bayer, I.S.; Athanassiou, A. Environmentally Benign Production of Stretchable and Robust Superhydrophobic Silicone Monoliths. *ACS Appl. Mater. Interfaces* **2018**, *10*, 2907–2917. [CrossRef]

100. Khorsand, S.; Raeissi, K.; Ashrafizadeh, F.; Arenas, M.A. Super-Hydrophobic Nickel-Cobalt Alloy Coating with Micro-Nano Flower-Like Structure. *Chem. Eng. J.* **2015**, *273*, 638–646. [CrossRef]

101. Sun, Y.; Sun, Q.; Huang, H.; Aguila, B.; Niu, Z.; Perman, J.A.; Ma, S. A Molecular-Level Superhydrophobic External Surface to Improve the Stability of Metal-Organic Frameworks. *J. Mater. Chem. A* **2017**, *5*, 18770–18776. [CrossRef]

102. Qian, Z.; Wang, S.; Ye, X.; Liu, Z.; Wu, Z. Corrosion Resistance and Wetting Properties of Silica-Based Superhydrophobic Coatings on AZ31B Mg Alloy Surfaces. *Appl. Surf. Sci.* **2018**, *453*, 1–10. [CrossRef]

103. Lv, D.; Ou, J.; Xue, M.; Wang, F. Stability and Corrosion Resistance of Superhydrophobic Surface on Oxidized Aluminum in NaCl Aqueous Solution. *Appl. Surf. Sci.* **2015**, *333*, 163–169. [CrossRef]

104. Li, L.; Bai, Y.; Li, L.; Wang, S.; Zhang, T. A Superhydrophobic Smart Coating for Flexible and Wearable Sensing Electronics. *Adv. Mater.* **2017**, *29*, 1702517. [CrossRef]

105. Liu, C.; Liu, Q.; Huang, A. A Superhydrophobic Zeolitic Imidazolate Framework (ZIF-90) with High Steam Stability for Efficient Recovery of Bioalcohols. *Chem. Commun.* **2016**, *52*, 3400–3402. [CrossRef]

106. Xu, L.; Zhu, D.; Lu, X.; Lu, Q. Transparent, Thermally and Mechanically Stable Superhydrophobic Coating Prepared by an Electrochemical Template Strategy. *J. Mater. Chem. A* **2015**, *3*, 3801–3807. [CrossRef]

107. Feng, L.; Yang, M.; Shi, X.; Liu, Y.; Wang, Y.; Qiang, X. Copper-Based Superhydrophobic Materials with Long-Term Durability, Stability, Regenerability, and Self-Cleaning Property. *Colloids Surf. A Physicochem. Eng. Asp.* **2016**, *508*, 39–47. [CrossRef]

108. Ke, X.; Zhou, X.; Hao, G.; Xiao, L.; Liu, J.; Jiang, W. Rapid Fabrication of Superhydrophobic Al/Fe₂O₃ Nanothermite Film with Excellent Energy-Release Characteristics and Long-Term Storage Stability. *Appl. Surf. Sci.* **2017**, *407*, 137–144. [CrossRef]

109. Zhao, Q.; Tang, T.; Wang, F. Fabrication of Superhydrophobic AA5052 Aluminum Alloy Surface with Improved Corrosion Resistance and Self Cleaning Property. *Coatings* **2018**, *8*, 390. [CrossRef]

110. Zhang, B.; Xu, W.; Zhu, Q.; Yuan, S.; Li, Y. Lotus-Inspired Multiscale Superhydrophobic AA5083 Resisting Surface Contamination and Marine Corrosion Attack. *Materials* **2019**, *12*, 1592. [CrossRef]

111. Zhang, F.; Ju, P.; Pan, M.; Zhang, D.; Huang, Y.; Li, G.; Li, X. Self-Healing Mechanisms in Smart Protective Coatings: A Review. *Corros. Sci.* **2018**, *144*, 74–88. [CrossRef]

112. Leal, D.A.; Riegel-Vidotti, I.C.; Ferreira, M.G.S.; Marino, C.E.B. Smart Coating Based on Double Stimuli-Responsive Microcapsules Containing Linseed Oil and Benzotriazole for Active Corrosion Protection. *Corros. Sci.* **2018**, *130*, 56–63. [CrossRef]

113. Li, J.; Feng, Q.; Cui, J.; Yuan, Q.; Qiu, H.; Gao, S.; Yang, J. Self-Assembled Graphene Oxide Microcapsules in Pickering Emulsions for Self-Healing Waterborne Polyurethane Coatings. *Compos. Sci. Technol.* **2017**, *151*, 282–290. [CrossRef]

114. Cho, S.H.; White, S.R.; Braun, P.V. Self-Healing Polymer Coatings. *Adv. Mater.* **2009**, *21*, 645–649. [CrossRef]

115. Wong, T.S.; Kang, S.H.; Tang, S.K.; Smythe, E.J.; Hatton, B.D.; Grinthal, A.; Aizenberg, J. Bioinspired Self-Repairing Slippery Surfaces with Pressure-Stable Omniphobicity. *Nature* **2011**, *477*, 443–447. [CrossRef]

coatings

MDPI

Article

Synthesis and Properties of Electrodeposited Ni–Co/WS$_2$ Nanocomposite Coatings

Yang He [1,*], Shuncai Wang [2], Wanting Sun [3], Philippa A.S. Reed [4] and Frank C. Walsh [2]

[1]   Centre for Composite Materials and Structures, Harbin Institute of Technology, Harbin 150080, China
[2]   National Centre for Advanced Tribology at Southampton (nCATS), University of Southampton,
      Southampton SO17 1BJ, UK; wangs@soton.ac.uk (S.W.); F.C.Walsh@soton.ac.uk (F.C.W.)
[3]   School of Materials Science and Engineering, Harbin Institute of Technology, Harbin 150001, China;
      sunwt_hit@126.com
[4]   Engineering Materials and Surface Engineering, University of Southampton, Southampton SO17 1BJ, UK;
      P.A.Reed@soton.ac.uk
*   Correspondence: yang.he@hit.edu.cn

Received: 17 January 2019; Accepted: 13 February 2019; Published: 25 February 2019

Abstract: Ni–Co coatings have gained widespread attention due to their potential in replacing hard chromium deposits (which have traditionally utilized toxic and corrosive chromic acid baths). A major challenge is to lower the high coefficient of friction of coated surfaces against steel, under dry sliding conditions. In this research, low friction Ni–Co/WS$_2$ nanocomposite coatings have been prepared by a convenient, one-pot electrodeposition from aqueous Ni–Co plating baths containing WS$_2$ particles. The embedment of the WS$_2$ lubricants is found to reduce the friction coefficient of coating significantly, and an ultra-low friction coefficient of 0.16 is obtained for the coating having a WS$_2$ content of 7.1 wt.%. Morphology and composition characterization of wear tracks reveal that the formation of a WS$_2$-rich lubricating tribofilm on the contact surfaces is beneficial to a low friction coefficient and good oxidation resistance. The wettability of electrodeposited coatings was also investigated. Compared to pure Ni-Co coating, the Ni–Co/7.1 wt.% WS$_2$ coating has an excellent hydrophobicity with a high water contact angle (WCA) of 157°, due to a rough surface with dual scale protrusions and the low surface energy of WS$_2$.

Keywords: Ni–Co; WS$_2$; hydrophobicity; low friction; nanocomposite

1. Introduction

As one of the most promising alternatives for replacement of hard-chromium coatings, Ni–Co coatings have numerous advantages, including high hardness, good adhesion, superior wear and corrosion resistance. These deposits have been widely used in tribological applications [1–3]. A favored method for the preparation of Ni–Co alloy/composite coatings is electrodeposition, which is facile, reproducible, readily controlled and suitable for industrial scale-up [4]. Over the last three decades, considerable research studies have focused on tailoring the microstructure and tribological properties of Ni–Co coating via control of the electrolyte composition, electrodeposition conditions and heat treatment temperature [5–8].

A coating offering low friction against a steel counterpart can reduce the emission of long-lived greenhouse gases, including carbon dioxide (CO$_2$), nitrogen oxides (N$_2$O and NO$_2$) and methane (CH$_4$). The attempt to produce such a coating has attracted increasing attention in both fundamental research and industrial practice. Ni based metallic coatings, however, possess a range of high friction coefficients (in the range of 0.4 to 0.7) against steel under dry friction conditions, which cannot fulfil economic expectations and meet the rising demands for environmental protection. Composite coatings of solid lubricants in a metallic matrix, for example, CNTs [9,10], graphite [11,12], MoS$_2$ [13], WS$_2$ [14]

and reduced graphene oxide [15], have been widely developed to provide enhanced tribological properties (i.e., low friction, chemical inertness, good wear and corrosion resistance).

As a distinct solid lubricant, WS_2 has a very low coefficient of friction (approx. 0.01) owing to the facile shear of weak interlayer bonds (bound by van der Waal's forces) between layers. Furthermore, it has a good thermal stability up to 594 °C. These characteristics make WS_2 very suitable as a friction modifier in metallic coatings. Metal-WS_2 composite coatings with low friction coefficient and/or good wear resistance have been intensively prepared via electrodeposition over recent years. Tenne et al. [14] electrodeposited WS_2 particle impregnated Ni films on archwires, which showed a reduction in frictional force up to 60% in comparison to uncoated archwires. García-Lecina et al. [16] reported an electrodeposited Ni–WS_2 coating with a stable friction coefficient of 0.4 against steel. Tudela et al. [17] further improved the mechanical strength of the electrodeposited Ni–WS_2 coating to enhance bath agitation. Das et al. [18] demonstrated that pulse current electrodeposited Ni–WS_2 coating showed a very low friction coefficient of 0.12. In a previous study, we investigated the self-lubricating properties of an electrodeposited Ni–P/WS_2 composite coating [19]. Research on the promising combination of ceramic WS_2 lubricants and Ni–Co coatings is very limited.

Superhydrophobic coatings have wide potential applications in, e.g., self-cleaning, anti-icing, oil/water separation, anti-corrosion, anti-fouling, energy saving, and medical implants. Recently, superhydrophobic WS_2 composite surfaces including Ni/WS_2 [20] and Ni–P/WS_2 [19] have been readily obtained by electrodeposition without the need for expensive equipment or a specialized reaction environment.

In this study, we present work on the fabrication of low friction Ni–Co/WS_2 composite coatings via facile, one-pot electrodeposition in a modified Watts Nickel bath. The variations of microstructure, tribological and hydrophobic properties, which are dependent on the WS_2 content of the coating, are systematically investigated. In sliding friction studies, the wear tracks are thoroughly characterized and the self-lubricating mechanism is discussed in detail.

2. Experimental Details

2.1. Sample Preparation

Analytical reagents and distilled water were used to prepare the plating solution. The Ni–Co/WS_2 composite coating was deposited from an electrolyte bath consisting of $NiSO_4 \cdot 6H_2O$ (200 g L^{-1}), $NiCl_2 \cdot 6H_2O$ (40 g L^{-1}), $CoSO_4 \cdot 7H_2O$ (40 g L^{-1}), boric acid (30 g L^{-1}), saccharin (2 g L^{-1}), cetyltrimethylammonium bromide (CTAB) (0.1 g L^{-1}) and WS_2 particles (1, 5, 10, 15, 25 g L^{-1}). The pH value of the solution was maintained at 4. To improve the dispersion, the solution was placed in an ultrasonic water bath for 15 min before electrodeposition.

A TTi QL355T power station (Cambridge, UK) was chosen as the power source. The coatings were deposited at 3 A dm^{-2} for 45 min. The bath temperature was maintained at 45 °C by a Grant LTD6G water bath (Cambridge, UK). The electrolyte was continuously stirred by a PTFE-coated magnetic stirrer bar (6 mm diameter × 30 mm length) at 1.67 s^{-1}. The anode was a Ni sheet (purity 99.5%) with a thickness of 1 mm, supplied by Goodman Alloys Ltd. (Yorkshire, England). The cathode substrate was 3 mm thick AISI 1020 mild steel supplied by Goodman Alloys Ltd. (Yorkshire, UK) with a hardness of 150 HV. Both the anode and cathode were cut to a size of 80 mm × 10 mm. Before electrodeposition, the surfaces were mechanically polished with 400, 800, and 1200 grade waterproof abrasive paper sequentially and then ultrasonically cleaned with acetone to remove contamination. During deposition, the cathode was held vertical and parallel to the anode with an interelectrode gap of 25 mm.

2.2. Characterization

Linear sweep voltammetry (LSV) measurements were performed with an Autolab PGSTA30 system in conjunction with a conventional, three-electrode glass cell. The working electrode was a rectangular mild steel plate of dimensions 10 and 40 mm, and the counter electrode was a platinum mesh electrode of similar dimensions. All LSV scans were recorded at a scan rate of 10 mV s^{-1} over the potential range of −0.5 V to −1.2 V vs. saturated calomel electrode (SCE).

The surface morphologies of as-deposited coatings and wear tracks were imaged using a JEOL JSM 6500 SEM (Tokyo, Japan), using an applied voltage of 15 kV and a working distance of 10 mm. Using integrated energy dispersive X-ray spectroscopy (Oxford Instruments, Oxford, UK), EDS, quantitative elemental information on samples was examined by the characteristic X-ray emission. The phase structures of the composite coatings were analyzed using a Bruker GADDS diffractometer with Cu Kα radiation, scanned at 0.02 deg s^{-1} over the 2θ range from 10° to 90°.

The friction and wear characteristics of the as obtained coatings were assessed using a reciprocating tribometer (Plint and Parners, Ltd., Wokingham, UK, model TE-77). The experiments were conducted at a constant temperature of 25 °C, in the dry sliding condition and at 40% relative humidity. AISI-52100 steel balls (diameter: 6 mm, hardness: 700 HV, Ra: 0.12 μm) were used as the sliding counter faces. The tests were carried out under a sliding frequency of 1 Hz and a sliding stroke of 2.69 mm for 1000 s. The normal contact load is 5 N and the frictional force is recorded automatically by a piezoelectric transducer. Friction tests on samples were repeated twice to ensure reliability.

Water contact angles (WCA) were determined using a commercial instrument Drop Shape Analysis system (KRÜSS, Hamburg, Germany, model DSA100) with a computer-controlled liquid dispensing system. The values of WCA are taken as an average of five measurements made with 6 μL distilled water droplets.

3. Results and Discussions

3.1. Linear Sweep Voltammetric (LSV) Analysis

The LSV method was utilized to analyze the mechanism of the coating formation on substrate. Negative direction, cathodic sweep was performed in the range from 500 to −1300 mV which was selected on the basis of the onset of alloy deposition. The cathodic LSV curve in Figure 1 shows that the reduction current density rises quickly as the potential decreases below −0.93 V, which corresponds to rapid simultaneous deposition of cobalt and nickel. After addition of WS$_2$ particles in the electrolyte, the LSV curve shows progressive polarization towards a more negative potential value of −1.01 V although the slope of the curve remains almost unchanged. The delay in electrodeposition is in agreement with results in Ni–Co/C nanotubes reported by Liu et al. [21], and can be attributed to the hindrance effect of WS$_2$ particles adsorbed on the electrode surface. The Ni–Co/WS$_2$ plating system with CTAB addition exhibits polarization towards a more positive potential value. This is due to the adsorption of CTAB cationic surfactant molecules on particles, accelerating the codeposition of WS$_2$ particles into the Ni–Co matrix.

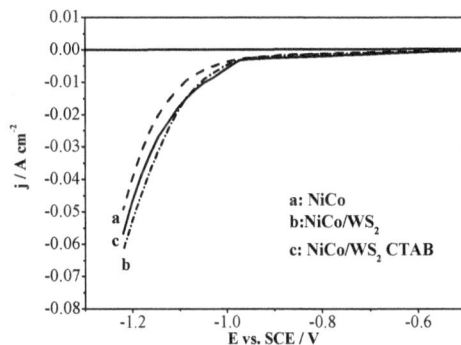

Figure 1. The current density-cathodic potential curves of (state electrode reactions) in Ni–Co and Ni–Co/WS$_2$ electrolytes.

3.2. Morphology and Structure

Figure 2 shows the WS_2 platelets with an irregular shape, which have diameters in the range of 80–240 nm. Figure 3 shows the surface morphologies of Ni–Co coating and Ni–Co/WS_2 coatings deposited from baths with different WS_2 concentrations. Compared to the flat morphology of the Ni–Co coating, Ni–Co/WS_2 coatings deposited at a WS_2 concentration of 1 g L^{-1} have a rough surface decorated with nodular protrusions. The density and diameter of the nodular protrusions increased at higher WS_2 concentration in the bath. During electrodeposition, Ni ions are preferably reduced around WS_2 semi conductive particles at the electrode, resulting in a high localized current distribution and the growth of Ni–Co/WS_2 bulges on the deposit surface.

Figure 2. (**a**) SEM image and (**b**) TEM image of as-received WS_2 particles.

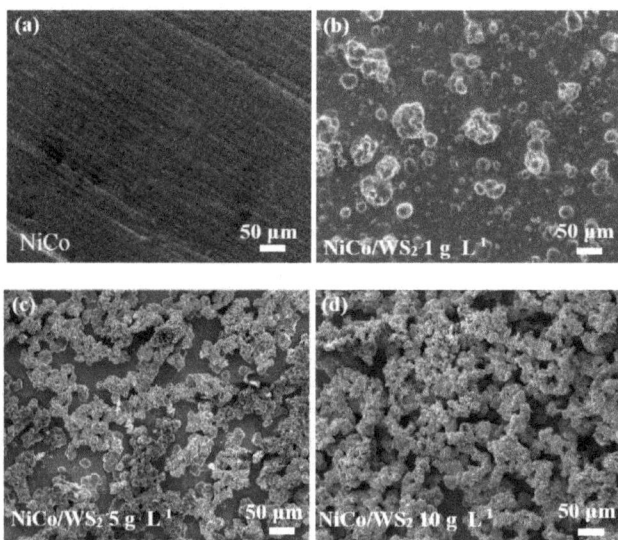

Figure 3. *Cont.*

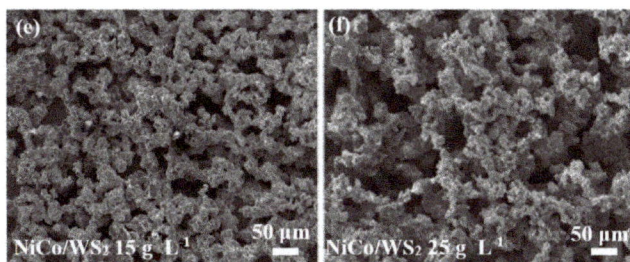

Figure 3. SEM images of the Ni-Co/WS$_2$ coatings deposited from solutions having controlled WS$_2$ concentrations: (**a**) 0 g L^{-1}, (**b**) 1 g L^{-1}, (**c**) 5 g L^{-1}, (**d**) 10 g L^{-1}, (**e**) 15 g L^{-1} and (**f**) 25 g L^{-1}.

The XRD patterns of WS$_2$ particles, Ni–Co and Ni–Co/WS$_2$ coatings are characterized and summarized in Figure 4. WS$_2$ particles show diffraction peaks at 14°, 33°, 40°, 49° and 58°, which are consistent with the (002), (100), (103), (106) and (110) orientations of hexagonal WS$_2$. Both Ni–Co and Ni–Co/WS$_2$ samples exhibit distinct (0002) hcp/(111) fcc growth orientation with an intense peak at $2\theta \approx 44°$. The Ni–Co/WS$_2$ sample also has an obvious peak at $2\theta = 14°$, confirming the successful inclusion of WS$_2$ in the coating. Using the Scherrer equation and observed diffraction peak widths, the mean grain sizes of Ni–Co/WS$_2$ 7.1 wt.% deposits can be calculated as 14 ± 2 nm.

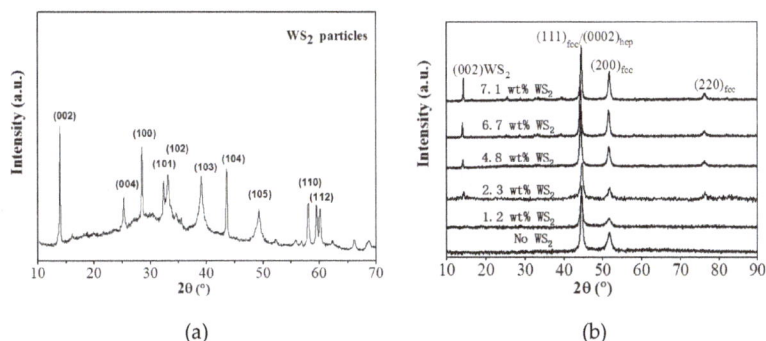

Figure 4. XRD patterns of (**a**) WS$_2$ particles and (**b**) Ni-Co/WS$_2$ samples with different WS$_2$ contents.

3.3. Compositional Analysis

The composition of electrodeposited coatings was examined by EDS. It can be seen in Figure 5 that the obtained coatings show a cobalt percentage in the range 35%–46%, which is much higher than the percentage of Co in the solutions. The trend is in consistence with the results in other Ni–Co electrodeposition systems [22]. The increment of the WS$_2$ particles in bath is found to lower the Co/Ni + Co ratio in coating, which is related to the electrochemical polarization caused by particle adsorption on the electrode [23].

Figure 5b shows that the weight percentage of WS$_2$ particles in coating increases with the increasing of the WS$_2$ concentration in bath, reaching a maximum WS$_2$ content of 7.1 wt.% at a particle concentration of 15 g L^{-1}. There is an optimal particle concentration in this experiment, as a further increase of WS$_2$ concentration leads to particle agglomeration and sedimentation in solution.

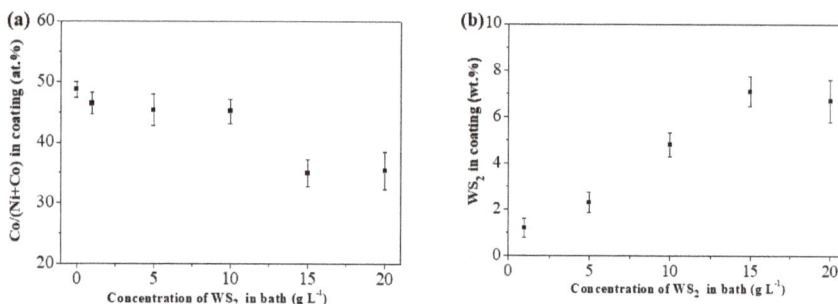

Figure 5. (a) The mol ratio of Co/(Co + Ni) and (b) the content of WS_2 particles in coating versus the concentration of WS_2 in solution.

3.4. Tribological Performance

Figure 6 shows the coefficient of friction (COF) versus time for samples having different WS_2 contents. The COF of the Ni–Co coating fluctuates within the range of 0.45 to 0.55 during the friction test. In contrast, the Ni–Co/WS_2 1.2 wt.% sample exhibits a lower COF of 0.15 in the initial 300 s, but followed by a slow rise to the range of 0.45–0.55 that is close to the COF of pure Ni–Co coating. Notably, the Ni–Co/WS_2 7.1 wt.% sample displays a stable low COF value of 0.16 for 1000 s, indicating the effective role of WS_2 lubricants on reduction of friction.

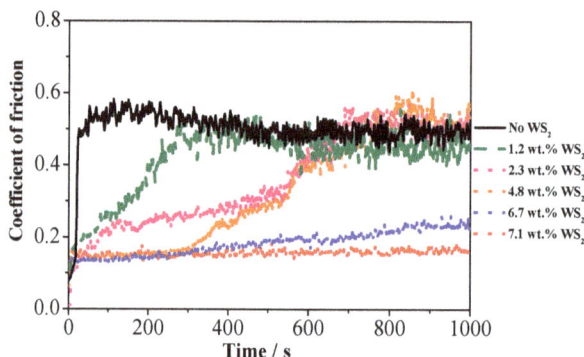

Figure 6. Relationship between friction coefficient and sliding time for the Ni–Co/WS_2 with different WS_2 contents.

The surface morphology and composition of wear tracks were examined by SEM and EDS respectively. Scoring wear can be found on the pure Ni–Co coating after wear test of 1000 s, as shown in Figure 7. The Ni–Co/WS_2 1.2 wt.% composite coating presents a similar wear track surrounded by lots of debris particles. It is worth noting that the surface protrusions of the electrodeposited coating are fragile and easily fractured during the friction process. With the increasing of WS_2 content in coating the wear gradually became less severe and finally changed to galling in the case of Ni–Co/WS_2 coatings with WS_2 content 2.3 wt.% and higher. The wear degree of the Ni-Co/WS_2 coating decreases with the increase of WS_2 content in coating. The Ni–Co/WS_2 7.1 wt.% sample shows only a narrow abrasive wear track with few debris particles. As listed in Table 1, the wear track of Ni–Co coating is composed of 22.1 wt.% Ni, 21.9 wt.% Co, 23.0 wt.% O and 33.0 wt.% Fe. The wear track of the Ni–Co/WS_2 7.1 wt.% sample contains a higher WS_2 content than that of the resting area on the coating, suggesting that WS_2 has a high adhesion ability to the sliding surface. The wear track of the Ni–Co/WS_2 7.1 wt.% sample shows a smaller oxygen content of 1.8 wt.% compared to that of pure

Ni–Co coating. On the one side, the release of WS$_2$ from the coating matrix reduces frictional heat on the mating surfaces, and on the other, the formation of compact WS$_2$ rich tribofilm separates the coating from the air. These two effects lead to a low degree of oxidation.

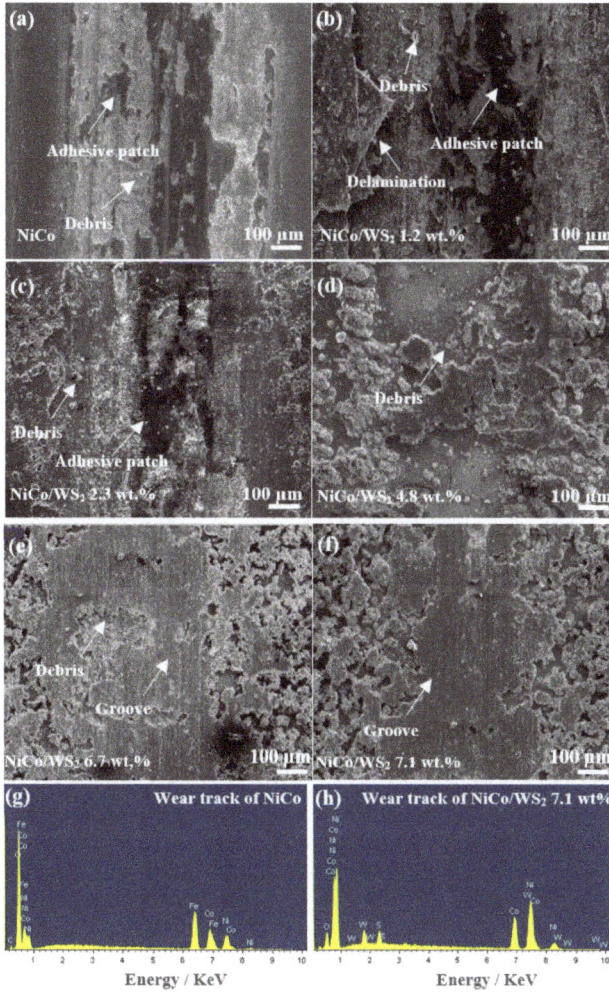

Figure 7. (**a–f**) SEM images of the wear tracks on various Ni–Co/WS$_2$ coatings; (**g–h**) EDS spectra of the wear tracks on Ni–Co coating and Ni–Co/WS$_2$ 7.1 wt.% coating.

Table 1. Composition of wear tracks on Ni-Co and Ni-Co/WS$_2$ coatings after friction testing.

Element (wt.%)	Ni-Co	Ni-Co/WS$_2$ 1.2 wt.%	Ni-Co/WS$_2$ 2.3 wt.%	Ni-Co/WS$_2$4.8 wt.%	Ni-Co/WS$_2$6.7 wt.%	Ni-Co/WS$_2$7.1 wt.%
Ni K	22.1	28.5	47.3	48.5	57.5	58.2
Co K	21.9	25.7	40.9	41.6	31.6	31.3
W M	–	0.9	1.5	4.4	5.8	6.9
S K	–	0.3	0.5	1.5	2.1	1.9
O K	23.0	17.1	4.8	4.0	3.0	1.8
Fe K	33.0	27.5	5.1	–	–	–

Figure 8 shows a comparison between Ni–Co coating and Ni–Co/WS$_2$ coating in the wear volume in the friction test. Ni–Co coating has a wear volume of 9.5×10^{-3} mm^{-3}. The wear volume of the Ni–Co/WS$_2$ 1.2 wt.% coating is as high as 2.0×10^{-2} mm^{-3} which is related to delamination failure. It is evident that the amount of codeposited WS$_2$ particles is important for improving the wear resistance of the Ni-Co coating when sliding against the steel ball, as indicated by lower volume of 6.0×10^{-3} mm^{-3} and 5.5×10^{-3} mm^{-3} for the Ni–Co/WS$_2$ 6.7 wt.% and Ni–Co/WS$_2$ 7.1 wt.% coatings.

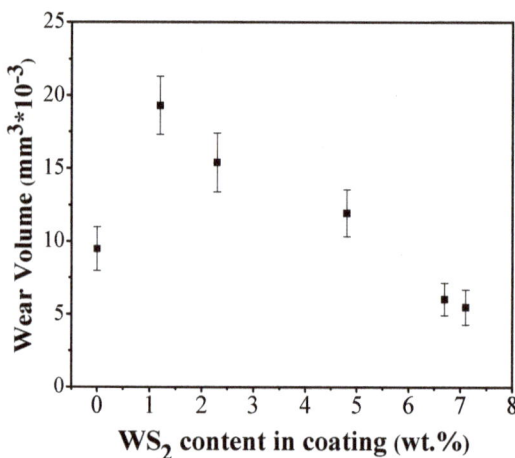

Figure 8. Comparison of wear volumes of Ni–Co coating and Ni–Co/WS$_2$ coatings with different WS$_2$ contents.

A proper run-in period is vital to the friction and wear performance of the electrodeposited coatings. Due to lack of sufficient solid lubricants, Ni–Co, Ni–Co/WS$_2$ 1.2 wt.% and Ni–Co/WS$_2$ 2.3 wt.% are not well run-in when sliding against the counter steel ball. For the Ni–Co/WS$_2$ coatings with WS$_2$ content 6.7 wt.% and higher, solid lubricants are reserved within the Ni–Co matrix and gradually released to the friction surfaces for lubrication, resulting in fast run-in and low friction and fundamentally inhibiting excessive operating temperatures and scuffing damage especially at high speeds.

Figure 9 the surface morphologies of counter steel balls after sliding tests. A clear wear scar is observable on the wear track of the steel ball after sliding against Ni–Co coating. Lots of wear debris is observed on the counter ball after sliding against the Ni–Co/WS$_2$ 1.2 wt.% coating, which derives from the breakdown of the nodular microstructures in the coating. In contrast, there is a smooth wear track on the counterpart ball after friction against the Ni–Co/WS$_2$ 7.1 wt.% coatings, suggesting effective lubrication by WS$_2$ from the coating.

For the Ni–Co coating, the adhesion force between the counterpart ball and the Ni–Co alloy surface is strong; debris particles generated during the friction test are crushed and piled up along the wear scar, i.e., not over-rolled. The wear particles slide firmly on the surface and turn into incomplete tribofilm during a long sliding time.

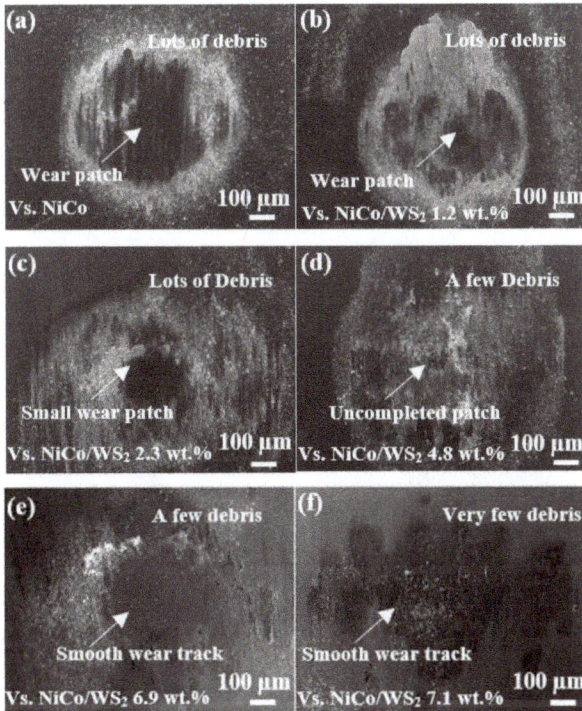

Figure 9. SEM images of counter balls after friction tests against (**a**) Ni–Co coating and Ni–Co/WS$_2$ coatings with different WS$_2$ contents: (**b**) 1.2 wt.%, (**c**) 2.3 wt.%, (**d**) 4.8 wt.%, (**e**) 6.7 wt.% and (**f**) 7.1 wt.%.

Due to the lamellar structure of WS$_2$, the inclusion of WS$_2$ particles into the Ni–Co coating significantly reduces the shear force (τ) causing a decrease in the friction coefficient, especially when a smooth lubricating layer is produced. The reconditioning and self-lubricating processes of Ni–Co/WS$_2$ coating is proposed as follows. Once friction starts, the movement of the counterpart ball will impose high shear stress on the contact area, resulting in severe deformation in the coating surface. Then the embedded WS$_2$ nanoparticles in the coating can release from their fixed positions to enter the sliding surfaces. Some WS$_2$ lamellae will be exfoliated into fine particles, which are easily accumulated in the wear crevices or attached to the metal surface to form a densely packed super-lubricating layer. The lubricating layer not only firmly attaches to the coating surface, but also transfers to the counterpart, thus contributing to the excellent tribological properties such as low friction coefficient, good wear resistance and high oxidation resistance under dry sliding conditions without oil or grease.

3.5. Wettability

The wetting properties of electrodeposited surfaces are determined by the water contact angle test. As seen in Figure 10, the pure Ni–Co coating has a low WCA of 85°. With the WS$_2$ content increasing, the WCA of the Ni–Co/WS$_2$ coating significantly increases. The Ni–Co/WS$_2$ 7.1 wt.% coating shows an excellent superhydrophobicity as indicated by a high WCA of 156.9 deg.

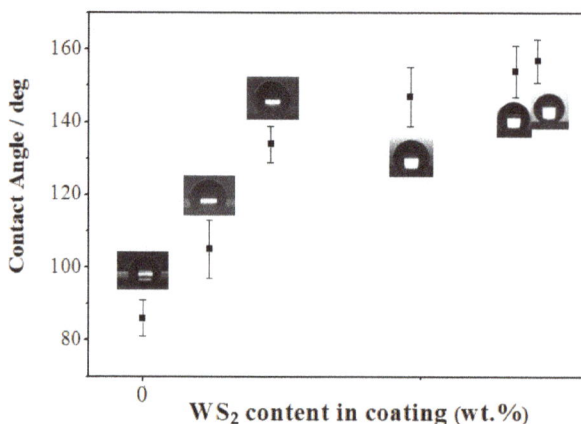

Figure 10. The variation in contact angle of water on the electrodeposited Ni–Co/WS$_2$ coatings against the concentrations of WS$_2$ in solution.

Figure 11 illustrates that the water droplet suspended on a needle tip is difficult to deposit on the coating even though it has been severely deformed. The water droplet rolls to one side when the needle tip approaches the sample surface, but can remain attached to the tip as the needle leaves the sample. This phenomenon indicates that the electrodeposited Ni–Co/WS$_2$ coating has a very low surface energy.

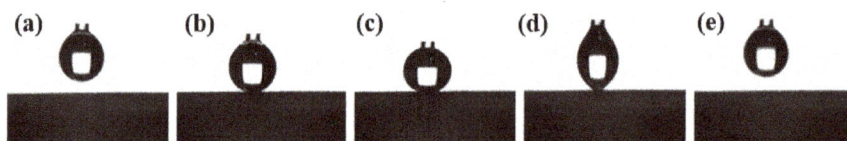

Figure 11. (**a**–**e**) Photographs illustrating water droplet does not adhere to the Ni–Co/WS$_2$ 7.1 wt.% surface: (**a**) a needle with suspended water droplet is set on the top of the Ni–Co/WS$_2$ coating; (**b**) the water droplet touches the surface of the coating with a low approaching velocity; (**c**) the water droplet become rounder as the needle further approach; (**d**) the water droplet is deformed heavily when the needle is separating from surface; (**e**) the water droplet is away from the coating after separation.

As well known, surface wettability is determined by chemical composition and surface structure, and can be described by either the Wenzel [24] or Cassie–Baxter model [25]. The Ni–Co surface has a WCA below 90° and is intrinsically hydrophilic. According to Wenzel's model, the wettability of intrinsic hydrophilic surface should be enhanced as the roughness increases, which is obviously not applicable for the electrodeposited Ni–Co/WS$_2$ coatings. Cassie–Baxter's model is commonly used to predict the hydrophobicity when air is trapped between the solid surface and the liquid droplet. The Ni–Co/WS$_2$ surfaces have a double-rough structure consisting of submicron asperities on micron-scale protrusions as illustrated in Figure 12; therefore, when these coatings are subjected to a liquid droplet, air bubbles are trapped and retained within asperities on the surface, maintaining the hydrophobic Cassie–Baxter state. The dual scale roughness has already been applied as the most commonly used criterion for superhydrophobic solid surface in the Cassie–Baxter state [26].

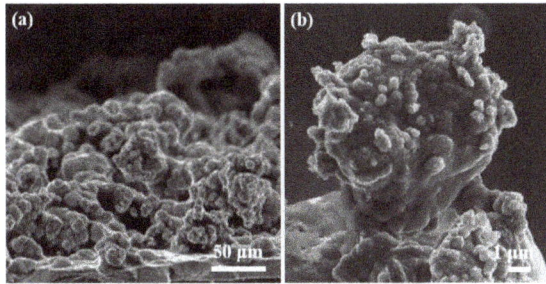

Figure 12. (**a**) Low- and (**b**) high- magnification SEM images of the cross-section of a Ni–Co/WS₂ 7.1 wt.% coating.

4. Conclusions

Ni–Co/WS$_2$ nanocomposite coatings have been successfully fabricated using a rapid one-pot electrodeposition route. The increase of the WS$_2$ content plays an important role in lowering the friction coefficient of the electrodeposited coating against steel to 0.16. The analysis of the morphology and composition of wear scars reveal that WS$_2$ strongly adheres to the sliding interfaces for effective lubrication rather than being squeezed out. Adequate levels of WS$_2$ in the coating could enable the formation of a compact, WS$_2$-rich tribofilm on the sliding interfaces that accounts for the ultra-low friction coefficient and high oxidation resistance. Moreover, the hierarchical rough Ni–Co/WS$_2$ composite coating exhibits excellent superhydrophobicity with a high WCA of 158°.

Author Contributions: Conceptualization, Y.H. and S.W.; Methodology, Y.H.; Investigation, Y.H. and W.S.; Formal Analysis, Y.H. and S.W.; Writing–Original Draft Preparation, Y.H.; Writing–Review and Editing, Y.H., S.W., W.S., P.A.S.R. and F.C.W.; Supervision, S.W., P.A.S.R. and F.C.W.; Project Administration, Y.H.; Funding Acquisition, Y.H.

Funding: This research was funded by National Natural Science Foundation of China (No. 11802077).

Conflicts of Interest: The authors declare no conflict of interest.

References

1. Tian, L.; Xu, J.; Xiao, S. The influence of pH and bath composition on the properties of Ni–Co coatings synthesized by electrodeposition. *Vacuum* **2011**, *86*, 27–33. [CrossRef]
2. Wang, L.; Gao, Y.; Xue, Q.; Liu, H.; Xu, T. Microstructure and tribological properties of electrodeposited Ni–Co alloy deposits. *Appl. Surf. Sci.* **2005**, *242*, 326–332. [CrossRef]
3. Go, E.; Ramirez, J.; Valle, E. Electrodeposition of Co–Ni alloys. *J. Appl. Electrochem.* **1998**, *28*, 71–79.
4. He, Y.; Wang, S.C.; Walsh, F.C.; Li, W.S.; He, L.; Reed, P.A.S. The monitoring of coating health by in situ luminescent layers. *RSC Adv.* **2015**, *5*, 42965–42970. [CrossRef]
5. Correia, A.N.; Machado, S.A.S. Electrodeposition and characterisation of thin layers of Ni–Co alloys obtained from dilute chloride baths. *Electrochim. Acta* **2000**, *45*, 1733–1740. [CrossRef]
6. Qiao, G.; Jing, T.; Wang, N.; Gao, Y.; Zhao, X.; Zhou, J.; Wang, W. High-speed jet electrodeposition and microstructure of nanocrystalline Ni–Co alloys. *Electrochim. Acta* **2005**, *51*, 85–92. [CrossRef]
7. Hibbard, G.D.; Aust, K.T.; Erb, U. Thermal stability of electrodeposited nanocrystalline Ni–Co alloys. *Mater. Sci. Eng. A* **2006**, *433*, 195–202. [CrossRef]
8. Ma, C.; Wang, S.C.; Walsh, F.C. Electrodeposition of nanocrystalline nickel–cobalt binary alloy coatings: A review. *Trans. IMF* **2015**, *93*, 104–112. [CrossRef]
9. Chen, W.X.; Tu, J.P.; Wang, L.Y.; Gan, H.Y.; Xu, Z.D.; Zhang, X.B. Tribological application of carbon nanotubes in a metal-based composite coating and composites. *Carbon* **2003**, *41*, 215–222. [CrossRef]
10. Akbarpour, M.R.; Alipour, S.; Safarzadeh, A.; Kim, H.S. Wear and friction behavior of self-lubricating hybrid Cu-(SiC + xCNT) composites. *Compos. Part B Eng.* **2019**, *158*, 92–101. [CrossRef]
11. Zhao, H.; Liu, L.; Hu, W.; Shen, B. Friction and wear behavior of Ni–graphite composites prepared by electroforming. *Mater. Des.* **2007**, *28*, 1374–1378. [CrossRef]

12. Sun, W.C.; Zhang, P.; Zhao, K.; Tian, M.M.; Wang, Y. Effect of graphite concentration on the friction and wear of Ni–Al$_2$O$_3$/graphite composite coatings by a combination of electrophoresis and electrodeposition. *Wear* **2015**, *342*, 172–180. [CrossRef]

13. He, Y.; Wang, S.C.; Walsh, F.C.; Chiu, Y.L.; Reed, P.A.S. Self-lubricating Ni-P-MoS$_2$ composite coatings. *Surf. Coat. Technol.* **2016**, *307*, 926–934. [CrossRef]

14. Redlich, M.; Gorodnev, A.; Feldman, Y.; Kaplanashiri, I.; Tenne, R.; Fleischer, N.; Genut, M.; Feuerstein, N. Friction reduction and wear resistance of electro-co-deposited inorganic fullerene-like WS$_2$ coating for improved stainless steel orthodontic wires. *J. Mater. Res.* **2008**, *23*, 2909–2915. [CrossRef]

15. Mai, Y.J.; Zhou, M.P.; Ling, H.J.; Chen, F.X.; Lian, W.Q.; Jie, X.H. Surfactant-free electrodeposition of reduced graphene oxide/copper composite coatings with enhanced wear resistance. *Appl. Surf. Sci.* **2018**, *433*, 232–239. [CrossRef]

16. García-Lecina, E.; García-Urrutia, I.; Díez, J.A.; Fornell, J.; Pellicer, E.; Sort, J. Codeposition of inorganic fullerene-like WS$_2$ nanoparticles in an electrodeposited nickel matrix under the influence of ultrasonic agitation. *Electrochim. Acta* **2013**, *114*, 859–867. [CrossRef]

17. Tudela, I.; Zhang, Y.; Pal, M.; Kerr, I.; Cobley, A.J. Ultrasound-assisted electrodeposition of thin nickel-based composite coatings with lubricant particles. *Surf. Coat. Technol.* **2015**, *276*, 89–105. [CrossRef]

18. Roy, D.; Das, A.K.; Saini, R.; Singh, P.K.; Kumar, P.; Hussain, M.; Mandal, A.; Dixit, A.R. Pulse current co-deposition of Ni–WS$_2$ nano-composite film for solid lubrication. *Mater. Manuf. Proc.* **2017**, *32*, 365–372. [CrossRef]

19. He, Y.; Sun, W.T.; Wang, S.C.; Reed, P.A.S.; Walsh, F.C. An electrodeposited Ni–P/WS$_2$ coating with combined super-hydrophobicity and self-lubricating properties. *Electrochim. Acta* **2017**, *245*, 872–882. [CrossRef]

20. Zhao, G.; Xue, Y.; Huang, Y.; Ye, Y.; Walsh, F.C.; Chen, J.; Wang, S. One-step electrodeposition of a self-cleaning and corrosion resistant Ni/WS$_2$ superhydrophobic surface. *RSC Adv.* **2016**, *6*, 439–443. [CrossRef]

21. Shi, L.; Sun, C.F.; Gao, P.; Zhou, F.; Liu, W.M. Electrodeposition and characterization of Ni–Co–carbon nanotubes composite coatings. *Surf. Coat. Technol.* **2006**, *200*, 4870–4875. [CrossRef]

22. Golodnitsky, D.; Yu, R.; Ulus, A. The role of anion additives in the electrodeposition of nickel–cobalt alloys from sulfamate electrolyte. *Electrochim. Acta* **2003**, *47*, 2707–2714. [CrossRef]

23. Gómez, E.; Pané, S.; Vallés, E. Electrodeposition of Co–Ni and Co–Ni–Cu systems in sulphate–citrate medium. *Electrochim. Acta* **2006**, *51*, 146–153. [CrossRef]

24. Wenzel, R.N. Resistance of solid surfaces to wetting by water. *Ind. Eng. Chem.* **1936**, *28*, 988–994. [CrossRef]

25. Cassie, A.B.D.; Baxter, S. Wettability of porous surfaces. *Trans. Faraday Soc.* **1944**, *40*, 546–551. [CrossRef]

26. Shirtcliffe, N.J.; McHale, G.; Newton, M.I.; Chabrol, G.; Perry, C.C. Dualscale roughness produces unusually water repellent surfaces. *Adv. Mater.* **2004**, *16*, 1929–1932. [CrossRef]

coatings

MDPI

Article

Fabrication of Superhydrophobic AA5052 Aluminum Alloy Surface with Improved Corrosion Resistance and Self Cleaning Property

Qian Zhao [1], Tiantian Tang [2] and Fang Wang [2,*]

1   Shaanxi Key Laboratory of Disaster Monitoring and Mechanism Simulation, Baoji University of Arts and Sciences, Baoji 721013, China; qianzhao@bjwlxy.cn
2   College of Chemistry & Pharmacy, Northwest A&F University, Yangling 712100, China; tangtiantian1020@foxmail.com
*   Correspondence: wangfang4070@nwsuaf.edu.cn; Tel.: +86-29-8709-2226; Fax: +86-29-8709-2082

Received: 10 September 2018; Accepted: 26 October 2018; Published: 31 October 2018

Abstract: The development of a self-cleaning and corrosion resistant superhydrophobic coating for aluminum alloy surfaces that is durable in aggressive conditions has attracted great interest in materials science. In the present study, a superphydrophobic film was fabricated on an AA5052 aluminum alloy surface by the electrodeposition of Ni–Co alloy coating, followed by modification with 6-(N-allyl-1,1,2,2-tetrahydro-perfluorodecyl) amino-1,3,5-triazine-2,4-dithiol monosodium (AF17N). The surface morphology and characteristics of the composite coatings were investigated by means of scanning electron microscopy (SEM), energy dispersive X-ray spectrum (EDS), atomic force microscope (AFM) and contact angle (CA). The corrosion resistance of the coatings was assessed by electrochemical tests. The results showed that the surface exhibited excellent superhydrophobicity and self-cleaning performance with a contact angle maintained at 160° after exposed to the atmosphere for 240 days. Moreover, the superhydrophobic coatings significantly improved the corrosion resistant performance of AA5052 aluminum alloy substrate in 3.5 wt.% NaCl solution.

Keywords: superhydrophobic surface; aluminum alloy; corrosion resistance; self-cleaning

---

1. Introduction

Aluminum and its alloys have widespread engineering applications owing to their high strength-to-density ratio, ductility, low weight and formability [1,2]. The wrought aluminum-zinc-magnesium-copper series alloys have been largely employed for production of the AA5052 chemical equipment, pressure vessels, food packaging, fan blades, coding utilities, automotive parts, etc. [3,4]. However, because the potential of aluminum alloys are more negative compared to other conventional metals, they are highly susceptible to corrosion, especially in moist environments, and undergo more rapid deterioration due to localized corrosion than their homogeneous counterparts. This has seriously limited their widespread applications [5]. Increasing the hydrophobicity of metal surfaces can reduce their interactions with corrosive media such as water, thereby enhancing their corrosion resistance [6]. Therefore, a self-cleaning superhydrophobic surface could be a potential solution to solve the functionality and aesthetic appearance problems caused by corrosion and contamination.

In recent years, superhydrophobic surfaces, which exhibit a water contact angle (CA) above 150° and a sliding angle (SA) below 10°, have aroused an enormous amount of interest in both fundamental research and potential applications because of their unique characteristics, such as self-cleaning [7], anti-icing [8,9], oil–water separation [10,11], antifouling property [12] and anticorrosion [13]. Inspired by biological materials in nature, such as lotus leaves [14], a variety of metallic surfaces with super-hydrophobic property have been fabricated through the combination of surface micro/nano structures and low surface energy materials [15,16].

Wettability on the surface of the material mainly depends on the surface chemical properties and surface microstructure [17]; thus, the improvement of the hydrophobicity of material surface is often achieved by changing the surface microscopic structure and lowering the surface energy [18,19]. Up to now, artificial super hydrophobic surfaces on aluminum and its alloy substrates have been created by various methods [20–22]. However, practical use is often interrupted by time-consuming processes, low mechanical strength or expensive cost. Additionally, most approaches involve the use of fluorinated compounds, which are costly and environmentally undesirable. Herein, a simple, environmental-friendly and cost-effective approach is still a much-needed study.

Triazinedithiol compounds, as a kind of environmentally friendly compounds, have attracted many researchers' attention during the past decade for their good properties such as high reactive, low cost, good adhesion and dielectric property on a variety of metal category [23–25]. Specifically, the triazinedithiol polymeric nanofilm exhibits excellent hydrophobicity and corrosion resistance in previous studies [26].

In the present paper, we reported a simple and efficient process for the construction of a superhydrophobic surface on an AA5052 aluminum alloy, wherein Ni–Co platings were electrodeposited on the pretreated AA5052 substrate surface to first form hierarchical micro/nano structures, which was then modified using an environmentally-friendly long-chain triazinedithiol compound (AF17N) with a low surface energy. Additionally, the surface morphology, self-cleaning characteristics and anticorrosion behavior of the obtained surface was investigated.

2. Materials and Methods

2.1. Materials

Aluminum alloys (AA5052, size $50 \times 20 \times 0.3$ mm^3, chemical composition: Cu: 0.1 wt.%, Si: 0.2 wt.%, Fe: 0.4 wt.%, Mn: 0.1 wt.%, Mg: 2.8 wt.%, Zn: 0.1 wt.%, Cr: 0.3 wt.%, other impurities 0.15 wt.%, and the remaining element Al) were used as substrates. 6-(N-allyl-1,1,2,2-tetrahydroperfluorodecyl) amino-1,3,5-triazine-2,4-dithiol monosodium (AF17N) was prepared by the reaction between 6-(N-allyl-1,1,2,2-tetrahydroperfluorodecyl)-amine-1,3,5-triazine-2,4-dichloride and NaSH, according to the method described previously in [27]. The structure of AF17N is shown in Figure 1. The nickel plate with a purity of 99.99 wt.% was used as the anode in the electrodeposition process. The other chemical reagents were obtained from Aladdin Reagent Database Inc., Shanghai, China. All regents were of analytical grade and deionized water was used for all of the experiments.

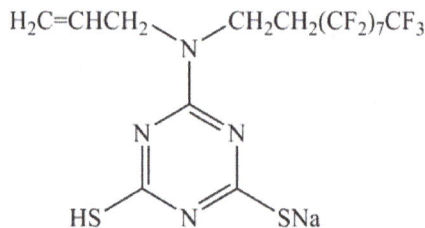

$$H_2C=CHCH_2 \diagdown \diagup CH_2CH_2(CF_2)_7CF_3$$

Figure 1. Structure of amino-1,3,5-triazine-2,4-dithiol monosodium (AF17N).

2.2. Preparation of the Superhydrophobic Coatings on AA5052 Surface

Synthesis of the superhydrophobic coatings on the AA5052 surface includes two steps: first, creation of a rough AA5052 surface by electrodepositing a Ni–Co plating; second, lowering the surface energy with AF17N.

The AA5052 substrate was initially degreased in an alkaline solution containing 10 g L^{-1} Na$_3$PO$_4$, 10 g L^{-1} Na$_2$CO$_3$ at 70 °C for 1.5 min, and then rinsed with deionized water. Subsequently, the AA5052 substrate was immersed in a solution (which contained 10 g L^{-1} NiCO$_3$·2NiO$_2$H$_2$·4H$_2$O, 5 g L^{-1} C$_6$H$_8$O$_7$·H$_2$O, 0.001 g L^{-1} (H$_4$N)$_2$S, 20 g L^{-1} NaH$_2$PO$_2$·H$_2$O and 30 mL L^{-1} NH$_3$·H$_2$O) for 40 min at 75 °C to form a electroless nickel coating [28]. Then, the samples were washed with deionized water, and last, dried in open air. The Ni–Co plating was electrodeposited using an electrochemical workstation (CHI 660C, CH Instrument, Shanghai, China) under direct current conditions. The electrodeposition was performed by a three-electrode cell, with the as-prepared samples with electroless nickel coating as a cathode, the saturated calomel electrode (SCE) as a reference electrode and the nickel plate was used as anode. The optimized bath composition and other parameters of electrodeposition Ni–Co alloy plating are given in Table 1.

Table 1. Bath compositions and operating conditions for electrodeposition of Ni–Co.

Compositions	Concentration (g L^{-1})	Conditions
NiCl$_2$·6H$_2$O	113	Current densities: 3 mA cm^{-2}
CoCl$_2$·6H$_2$O	8	pH: 3.6
C$_2$H$_8$N$_2$·2HCl	100	Temperature: 50 °C
H$_3$BO$_3$	15	Time: 360 s

Subsequently, the as-prepared AA5052 samples with Ni–Co alloy coatings were modified with 1 mM AF17N solution for 2 h at room temperature (25 °C ± 1 °C). Finally, the samples were rinsed with deionized water and dried in an oven (150 °C, 15 min) for further characterization.

2.3. Characterization

The water contact angles (CAs) and sliding angles (SAs) were measured with a telescopic goniometer (HARKE-SPCAX1). The volume of water drops was 3 μL. The values reported were the average values by measuring five different positions on each sample. A field emission scanning electron microscope (FESEM; HITACHI S-4800, Tokyo, Japan) and atomic force microscope (AFM; JSM-6360LV; JEOL, Tokyo, Japan) were used to characterize the surface morphologies. Scanning electron microscopy (SEM) with an energy dispersive X-ray spectrum (EDS) was used to characterize the surface chemical composition. The potentiodynamic polarization curves and electrochemical impedance spectroscopic (EIS) were performed on an electrochemical workstation (CHI 660C, CH Instrument, Shanghai, China) in a cell with 3.5 wt.% NaCl solution at room temperature (25 °C ± 1 °C). A three-electrode configuration was employed in all of the electrochemical tests, which consisted of the sample as the working electrode (1 cm^2), a graphite plate as the counter electrode and a saturated calomel electrode (SCE) as the reference electrode. The polarization curves were recorded with a sweep rate of 1 mV s^{-1}. EIS plots were performed in the frequency range between 100 kHz (high frequency area) and 10 mHz (low frequency area) with a sine-wave amplitude of 5 mV. The polarization curves and EIS spectra were fitted by using the CorrTest software. Cyclic voltammetry (CV) was conducted between the potential region from −0.7 V to 0.7 V at 20 mV s^{-1} in 0.1 M NaOH aqueous solution for three circles. All the electrochemical tests were normally repeated at least three times under the same conditions, indicating that they presented reasonable reproducibility.

3. Results

3.1. Surface Morphology and Chemical Compositions

The surface wettability of the as-obtained superhydrophobic surface was studied by CA and SA measurements. The images of the water droplets on various surfaces were shown in Figure 2. The bare aluminum surface exhibited a CA of 33.5° ± 1.6° (Figure 2a). After electrodeposition of Ni–Co, the sample surface exhibited superhydrophilicity with a CA of approximately 5° (Figure 2b).

Figure 2c shows that the static water droplet is in an approximately spherical shape on the level as-prepared surface and the water contact angle is as high as 160°, which is much higher than the bare aluminum alloy surface. Additionally, the advancing angle is about 161.3° and the receding angle is 158.7°. After modification with AF17N, the wettability changes from superhydrophilicity to superhydrophobicity. From Figure 2d, it can be observed that a 3 µL water droplet was bouncing on the surface and finally rolled off the nearly horizontal surface immediately without any adhesion, indicating that the superhydrophobic surface has an ultra-low sliding angle of about 1.0°. This means that the as-prepared superhydrophobic surface possesses excellent water-repellent property.

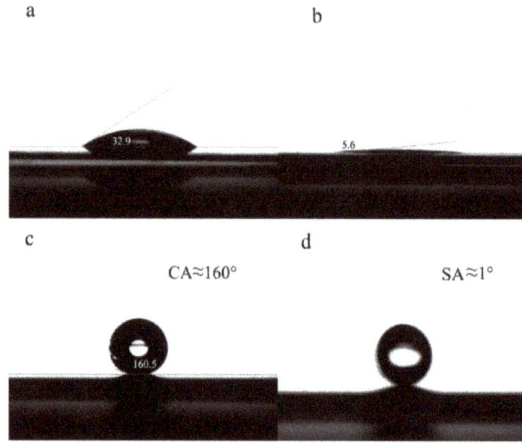

Figure 2. The contact angle (CA) images of (**a**) bare aluminum alloy; (**b**) electrodeposited Ni–Co plating; (**c**) superhydrophobic surface; (**d**) snapshot photograph of a water droplet rolling off on the tilted superhydrophobic surface.

This phenomenon could be explained by the Cassie–Baxter equation [29]:

$$cos\theta_r = f_1 cos\theta - f_2 \tag{1}$$

where f_1 and f_2 are fractional areas of the solid and air on the surface, respectively (i.e., $f_1 + f_2$ = 1); θ_r and θ are the water contact angles of the rough heterogeneous surfaces and smooth solid surfaces. According to Equation (1), with increasing the air surface fraction f_2, the solid surface fraction f_1 decreases, however, the contact angle of the rough surface increases. The air trapped in the grooves could reduce the contact area (f_1) between a water droplet and solid surface. Consequently, the hydrophobicity of the surface is enormously enhanced, and a high contact angle and low sliding angle would be obtained.

The SEM images of morphologies with different magnifications of the as-prepared superhydrophobic surface are given in Figure 3. It can be seen that the surface consists of two different microstructures. There are large amounts of small microparticles located on the electroless nickel coating (Figure 3a). Careful inspection of the surface reveals that the microparticles were cluster-like microclusters (Figure 3b) and the surface was completely and compactly covered with these cluster-like microclusters. From observing Figure 3c,d, it can be easily found that the cluster structures were composed of numerous irregular conelike structures with a width average length of 100–200 nm, indicating that the cluster-like structures have hierarchical micro/nano structures. More importantly, these hierarchical micro/nano structures can generate numerous grooves in which the air can be trapped easily, which can lead to the larger CA and smaller SA.

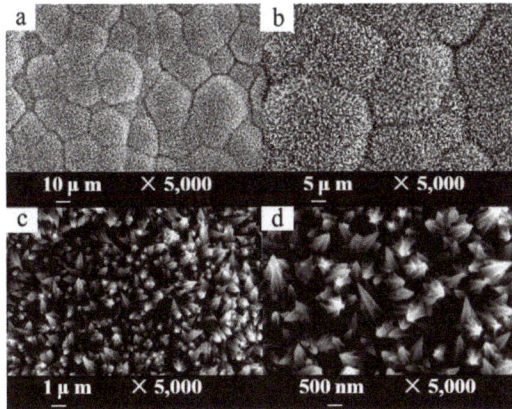

Figure 3. Scanning electron microscopy (SEM) images of the as-prepared superhydrophobic surface with different magnifications. (**a**) 5,000 times; (**b**) 10,000 times; (**c**) 50,000 times; (**d**) 100,000 times.

Chemical compositions of the surfaces were analyzed by using EDS (Figure 4). Figure 4a shows the EDS image of bare aluminum alloy. The EDS spectrum shows that no other evident peaks apart from that of Al, which indicates the surface is mainly composed of Al element. The EDS spectrum in Figure 4b reveals the additional presence of P and Ni elements, suggesting considerable changes in the composition of the aluminum alloy surface after the electroless nickel. In order to testify the AF17N, Figure 4c,d shows the EDS spectrum of the surface unmodified and modified by AF17N monomers after electrodeposition Ni–Co alloy plating. The EDS spectrum of as-perpared superhydrophobic aluminum surface in Figure 4d reveals the appearance of C, N, O, F, Al, P, Co and Ni elements. The presence of N and F elements are attributed to the AF17N, indicates that the surface has been covered with AF17N polymeric nanofilm after chemical modification, showing that the AF17N has been successfully adsorbed onto the surface (Figure 4d).

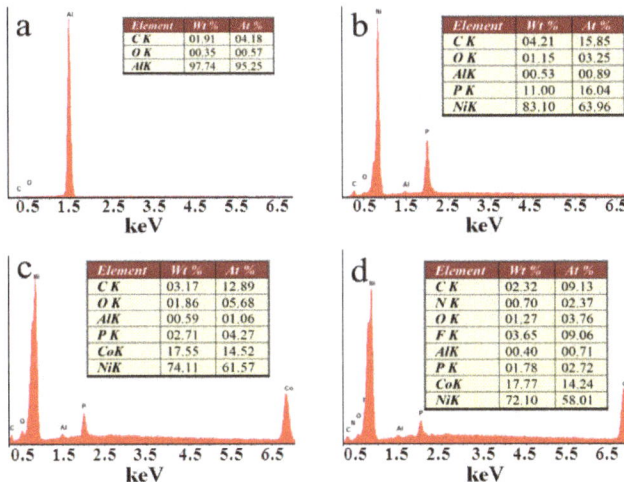

Figure 4. Energy dispersive X-ray spectrum (EDS) spectra and its corresponding element content. (**a**) bare aluminum alloy; (**b**) electroless Ni coating;(**c**) electrodeposited Ni–Co plating; (**d**) superhydrophobic surface after modified AF17N.

In order to find out the reason for the change of the wettability, SEM and AFM were applied. Figure 5 shows the SEM and AFM images of the aluminum alloy surface under different treatment conditions. Figure 5a,e demonstrate the micrographs of the bare aluminum alloy after pretreament. It is clear that the bare aluminum alloy surface is relatively flat without protuberance (surface roughness R_a is about 25.7 nm). According to Figure 5b,f, after electroless nickel coating, the aluminum alloy is covered with micro-sphere array structures, which have an average size of 500 nm in height and 2 μm in diameter. Simultaneously, the surface roughness distinctly increased (surface roughness R_a is about 76.7 nm). After electrodeposition of Ni–Co alloy plating, it is obvious that the nickel coating surface was completely and densely covered with enormous small particles. Moreover, the Ni–Co alloy plating has a frosting surface morphology compared with the nickel coating, which may be attributed to the smaller average size for the Ni–Co particles. It can be seen that the surface becomes quite rough (surface roughness R_a is about 92.5 nm) and develops mountain-like structure with the size of 864 nm in height (Figure 5g). The rough surface could then be changed from superhydrophilicity to superhydrophobic after modification by AF17N. It can be easily seen that the surface morphologies of the resulting superhydrophobic surface shown in Figure 5d,h are nearly the same as that of the Ni–Co alloy plating. This means that the surface morphologies of the Ni–Co alloy plating before and after modification with AF17N show hardly any change. This is because only a small amount of AF17N was absorbed onto the Ni–Co alloy plating. However, the surface roughness slightly decreased (surface roughness R_a is about 84.3 nm), owing to the AF17N might fill in some nano-scale pores of the rough surface (Figure 5h).

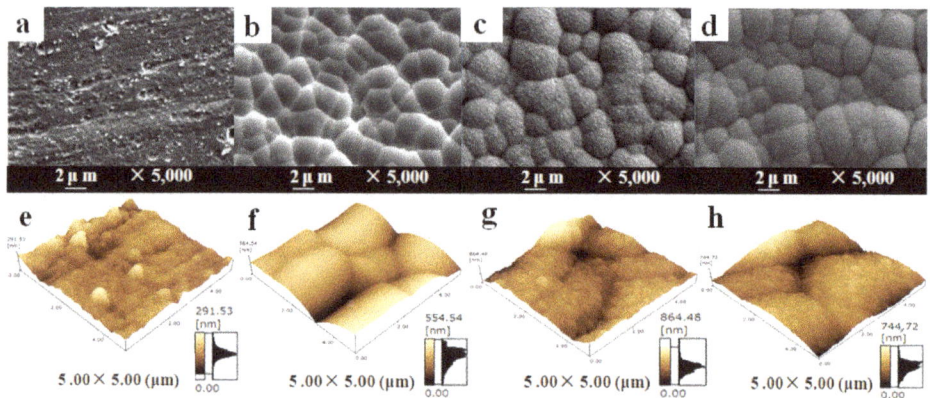

Figure 5. The SEM (**a–d**) and its corresponding atomic force microscope (AFM) (**e–h**) images. (**a–e**) bare aluminum alloy; (**b–f**) electroless Ni coating; (**c–f**) electrodeposited Ni–Co plating; (**d–h**) superhydrophobic surface.

3.2. Chemical Stability and Corrosion Resistance

To elucidate the stability of sample surfaces, the superhydrophobicity of the obtained surfaces were exposed to open air (room temperature of 24–26 °C and relative humidity of 40%–50%) and the contact angles were measured. Figure 6 shows the variation in the water contact angles of the as-prepared superhydrophobic surface with different exposure times. The water contact angles change from 151.3° ± 2.5° to 155.6° ± 2.1° after 4 weeks. When exposure was more than 16 weeks, the water contact angles increased, rather than decreased, from 155.6° to 160.0°. The results indicate that the as-obtained surface exhibit long-term stability; this is of great importance to the practical application of superhydrophobic surface.

Corrosion resistance ability was performed on the superhydrophobic AA5052 surface using the potentiodynamic polarization and electrochemical impedance spectroscopy (EIS) method. Figure 7 depicts the potentiodynamic polarization curves for the bare aluminum alloy substrate and the as-prepared superhydrophobic surface in neutral 3.5 wt.% NaCl solution. The relevant electrochemical parameters including corrosion potential and corrosion current density derived from polarization curves using the Tafel extrapolation method are listed in Table 2. Corrosion potential (E_{corr}), corrosion current density (I_{corr}), corrosion rate and protection efficiency (PE) are often applied to evaluate the corrosion protective property of the coatings. According to Table 2, the result clearly shows that the corrosion potential (E_{corr}) increase in positive direction from −0.69 V of the bare aluminum alloy to −0.28 V of the as-prepared superhydrophobic surface. Additionally, the corrosion current density (I_{corr}) decreased from 1.03×10^{-2} A cm^{-2} of bare aluminum alloy to 6.76×10^{-5} A cm^{-2} of as-prepared superhydrophobic surface. Moreover, its corrosion rate is merely 0.6% of the bare aluminum alloy.

The protection efficiency (PE) was calculated by using the expression:

$$PE(\%) = 100 \times [1 - (i/i_0)] \tag{2}$$

where i and i_0 are the corrosion current density of aluminum alloy with as-prepared super-hydrophobic surface and bare aluminum alloy, respectively. The protection efficiency calculated from potentiodynamic polarization data is found to be as high as 99.34%, demonstrating that the superhydrophobic surface significantly improves the corrosion resistance of aluminum alloys.

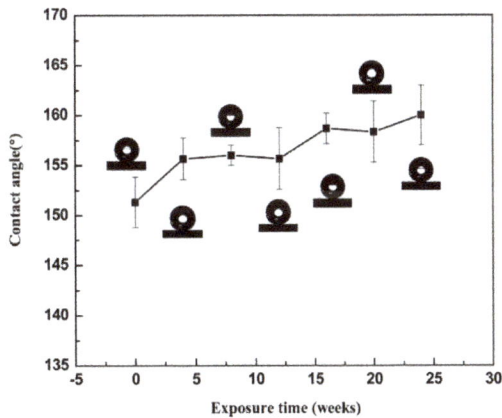

Figure 6. Variation in the water contact angles of samples surfaces with the exposure time.

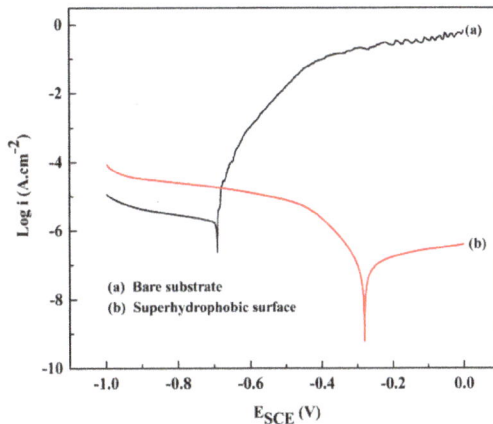

Figure 7. Potentiodynamic polarization curves of bare AA5052 aluminum alloy and superhydrophobic surface in 3.5 wt.% NaCl solution.

Table 2. Electrochemical parameters of potentiodynamic polarization curves.

Sample	E_{corr} (V)	I_{corr} (A cm^{-2})	Corrision Rate (mm a^{-1})	PE (%)
Bare aluminum alloy	−0.69	1.03×10^{-2}	1.12×10^{-1}	–
Superhydrophobic	−0.28	6.76×10^{-5}	7.39×10^{-4}	99.34

Figure 8 shows the Nyquist plots and Bode plots of the bare aluminum alloy and the as-prepared superhydrophobic surface. The results show quite different capacitive loops in the Nyquist plots. It is well known that the diameter of the capacitive loop represents the polarization resistance of the work electrode. As shown in Figure 8, the diameter of the as-prepared surface is obviously bigger than that of the bare substrate, which is attributed to a protective surface film of Ni–Co plating and AF17N polymeric nanofilms. Furthermore, we used the Bode plots to continue the investigation. From Figure 8, it can be observed that the impedance modulus |Z| of the as-prepared superhydrophobic surface is more than one order of magnitude higher than bare substrate, indicating that the superhydrophobic coating retards the formation of the corrosion products. Simultaneously, this result is consistent with that derived from the potentiodynamic polarization curves, showing that superhydrophobic surface can supply excellent corrosion protection for bare aluminum alloy in 3.5 wt.% NaCl solution.

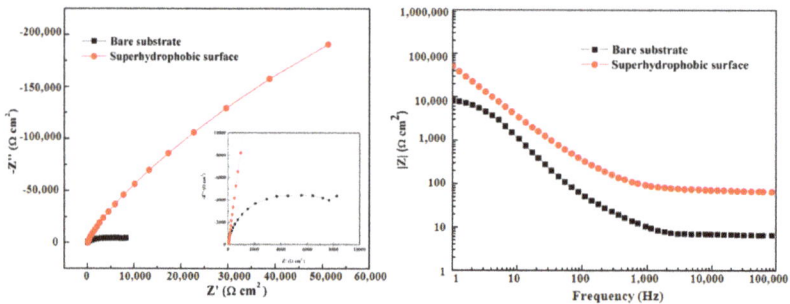

Figure 8. Nyquist plots and Bode plots of bare AA5052 aluminum alloy and superhydrophobic surface in 3.5 wt.% NaCl solution.

For better evaluation, the effect of AF17N polymeric nanofilm to the corrosion inhibition property of superhydrophobic AA5052 surface, the cyclic voltammetry (CV) curves of the bare aluminum alloy and the as-prepared superhydrophobic surface were measured in 0.1 mol L^{-1} NaOH aqueous solution to study the surface coverage of AF17N molecules (Figure 9). Noticeably, the surface covered with or without polymeric nanofilm has significant influence on the general shape of the cyclic voltammograms. For the surface without polymeric nanofilm, the activated anodic peaks at 0.37 V could be observed. It attributed to the formation of Ni(OH)$_2$ and Co(OH)$_2$ as a result of electro-oxidation of the Ni–Co plating. Cathodic peaks at 0.3 V was resulting from the reduction of Ni(OH)$_2$ and Co(OH)$_2$, respectively. Compared with the surface without polymeric nanofilm, the current and area of the superhydrophobic surface was much smaller. Thus, we can draw a conclusion that the changes observed in the curves of the superhydrophobic surface must result from specific interactions by AF17N monomer. These observations indicate that the oxidative reaction of superhydrophobic surface was highly blocked in alkaline aqueous, demonstrating the AF17N polymeric nanofilm is a good barrier to block charge or reactant ion transmission. It is expected that the resulting AF17N polymeric nanofilm supply excellent corrosion protection for bare aluminum alloy. Perfluoro alkyl chains in the polymer films are perpendicular to the substrate surface [30]. The hydrophobic group can isolate the water molecules effectively and provide good corrosion protection to aluminum alloy. In conclusion, superhydrophobic surfaces were fabricated with the –CF$_3$ group and –CF$_2$– group of low surface energy from the AF17N.

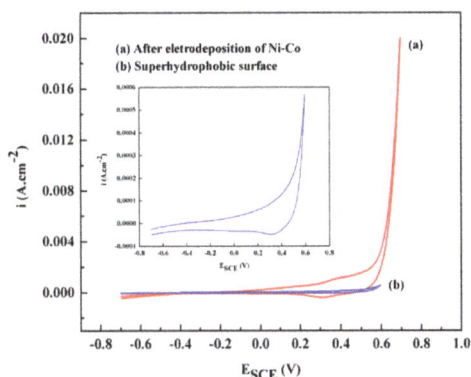

Figure 9. Cyclic voltammorgrams of bare AA5052 aluminum alloy and superhydrophobic surface in 0.1% NaOH solution.

3.3. Self-Cleaning Effect

The self-cleaning effect is an important character of superhydrophobic surfaces for their applications. The simulation process for the as-prepared superhydrophobic surface was carried out by deliberately sprinkling Al_2O_3 powder on the sample surface as a model dust contaminant. Figure 10 shows the evolution process of the self-cleaning effect, which was recorded by a digital camera at a speed of 60 frames per second. The sample with superhydrophobic surface was placed with a tilting angle of about 5° above horizontal and then a water droplet was dropped to the contaminant surface. When a water droplet was dropped to the sample surface, it rolled off immediately with removing all the Al_2O_3 powder along the path of the droplet. Amazingly, it was observed that the water droplet maintained its almost spherical shape. This result means that the as-prepared surface has an excellent self-cleaning effect due to the superior superhydrophobic property.

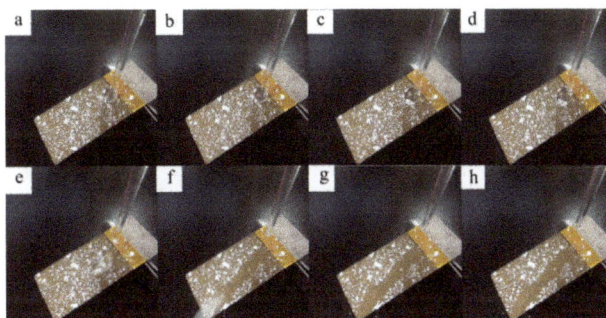

Figure 10. The time sequence of the self-cleaning process on the superhydrophobic surface at a sliding angle about 5°.

4. Conclusions

In summary, a functional superhydrophobic surface was successfully fabricated on AA5052 aluminum alloy by electrodeposition of Ni–Co films and surface modification with AF17N. The as-prepared superhydrophobic surface with hierarchical nano/micro structures has a high contact angle of 167.3° ± 1.3° and an ultra-low sliding angle of about 1°. It was found that the superhydrophobicity was governed by both geometrical microstructures and the surface chemical composition. Electrochemical measurements showed that the as-prepared sample possessed a better corrosion resistance than bare aluminum alloy, indicating that the superhydrophobic surface can effectively protect aluminum alloy from corrosion. Furthermore, the resulting superhydrophobic surface has good chemical stability and long-term durability as well as self-cleaning effect. This method is of great value for industrial preparation of superhydrophobic surfaces and it is also meaningful for extending the applications in other relevant engineered materials fields.

Author Contributions: Conceptualization and Methodology, F.W. and Q.Z.; Software, Validation, Investigation and Data Curation, Q.Z. and T.T.; Writing-Original Draft Preparation, Q.Z.; Writing-Review & Editing, F.W., Supervision, F.W.; Project Administration and Funding Acquisition, F.W. and Q.Z.

Funding: This research was supported by the Fundamental Research Funds for the Central Universities (No. QN2013085), the Natural Science Basic Research Plan in the Shaanxi Province of China (No. 2018JQ5191), the Young Talent fund of University Association for Science and Technology in Shaanxi, China (No. 20180419) and the Doctoral Scientific Research Foundation of Baoji University of Arts and Sciences (No. ZK2018044).

Conflicts of Interest: The authors declare no conflict of interest.

References

1. Loto, R.T.; Adeleke, A. Corrosion of aluminum alloy metal matrix composites in neutral chloride solutions. *J. Fail. Anal. Prev.* **2016**, *16*, 874–885. [CrossRef]
2. Quazi, M.M.; Fazal, M.A.; Haseeb, A.S.M.A.; Yusof, F.; Masjuki, H.H.; Arslan, A. Laser-based surface modifications of aluminum and its alloys. *Crit. Rev. Solid State Mater. Sci.* **2016**, *41*, 106–131. [CrossRef]
3. Barati Darband, G.; Aliofkhazraei, M.; Khorsand, S.; Sokhanvar, S.; Kaboli, A. Science and engineering of superhydrophobic surfaces: Review of corrosion resistance, chemical and mechanical stability. *Arab. J. Chem.* **2018**, in press. [CrossRef]
4. Gudić, S.; Vrsalović, L.; Kliškić, M.; Jerković, I.; Radonić, A.; Zekić, M. Corrosion inhibition of aa 5052 aluminium alloy in nacl solution by different types of honey. *Int. J. Electrochem. Sci.* **2016**, *11*, 998–1011.
5. Uwiringiyimana, E.; O'Donnell, P.S.; Joseph, I.V.; Adams, F.V. The effect of corrosion inhibitors on stainless steels and aluminium alloys: A review. *Afr. J. Pure Appl. Chem.* **2016**, *10*, 23–32.
6. Singh, B.P.; Jena, B.K.; Bhattacharjee, S.; Besra, L. Development of oxidation and corrosion resistance hydrophobic graphene oxide-polymer composite coating on copper. *Surf. Coat. Technol.* **2013**, *232*, 475–481. [CrossRef]
7. Asmone, A.S.; Chew, M.Y.L. An investigation of superhydrophobic self-cleaning applications on external building façade systems in the tropics. *J. Build. Eng.* **2018**, *17*, 167–173. [CrossRef]
8. Emelyanenko, A.M.; Boinovich, L.B.; Bezdomnikov, A.A.; Chulkova, E.V.; Emelyanenko, K.A. Reinforced superhydrophobic coating on silicone rubber for longstanding anti-icing performance in severe conditions. *Acs Appl. Mater. Interface* **2017**, *9*, 24210–24219. [CrossRef] [PubMed]
9. Nguyen, T.B.; Park, S.; Lim, H. Effects of morphology parameters on anti-icing performance in superhydrophobic surfaces. *Appl. Surf. Sci.* **2018**, *435*, 585–591. [CrossRef]
10. Wei, C.; Dai, F.; Lin, L.; An, Z.; He, Y.; Chen, X.; Chen, L.; Zhao, Y. Simplified and robust adhesive-free superhydrophobic SiO$_2$-decorated pvdf membranes for efficient oil/water separation. *J. Membr. Sci.* **2018**, *555*, 220–228. [CrossRef]
11. Li, H.; Zhao, X.; Wu, P.; Zhang, S.; Geng, B. Facile preparation of superhydrophobic and superoleophilic porous polymer membranes for oil/water separation from a polyarylester polydimethylsiloxane block copolymer. *J. Mater. Sci.* **2016**, *51*, 3211–3218. [CrossRef]
12. Yagüe, J.L.; Segadó, P.; Auset, M.; Borrós, S. Textured superhydrophobic films on copper prepared using solvent-free methods exhibiting antifouling properties. *Thin Solid Film* **2017**, *635*, 32–36. [CrossRef]
13. Mohamed, A.M.A.; Abdullah, A.M.; Younan, N.A. Corrosion behavior of superhydrophobic surfaces: A review. *Arab. J. Chem.* **2015**, *8*, 749–765. [CrossRef]
14. Yan, Y.Y.; Gao, N.; Barthlott, W. Mimicking natural superhydrophobic surfaces and grasping the wetting process: A review on recent progress in preparing superhydrophobic surfaces. *Adv. Coll. Interface Sci.* **2011**, *169*, 80–105. [CrossRef] [PubMed]
15. Liu, Y.; Liu, J.; Li, S.; Han, Z.; Yu, S.; Ren, L. Fabrication of biomimetic super-hydrophobic surface on aluminum alloy. *J. Mater. Sci.* **2014**, *49*, 1624–1629. [CrossRef]

16. Dey, S.; Chatterjee, S.; Singh, B.P.; Bhattacharjee, S.; Rout, T.K.; Sengupta, D.K.; Besra, L. Development of superhydrophobic corrosion resistance coating on mild steel by electrophoretic deposition. *Surf. Coat. Technol.* **2018**, *72*, 220–235. [CrossRef]

17. Qian, H.C.; Li, H.Y.; Zhang, D.W. Research progress of superhydrophobic surface technologies in the field of corrosion protection. *Surf. Technol.* **2015**, *3*, 15–24.

18. Hoshian, S.; Jokinen, V.; Somerkivi, V.; Lokanathan, A.R.; Franssila, S. Robust superhydrophobic silicon without a low surface-energy (hydrophobic) coating. *ACS Appl. Mater. Interface* **2015**, *7*, 941–950. [CrossRef] [PubMed]

19. Wang, J.; Liu, F.; Chen, H.; Chen, D. Superhydrophobic behavior achieved from hydrophilic surfaces. *Appl. Phys. Lett.* **2009**, *95*, 3063–3067. [CrossRef]

20. Forooshani, H.M.; Aliofkhazraei, M.; Rouhaghdam, A.S. Superhydrophobic aluminum surfaces by mechanical/chemical combined method and its corrosion behavior. *J. Taiwan Inst. Chem. Eng.* **2017**, *72*, 220–235. [CrossRef]

21. Xiong, J.; Sarkar, D.K.; Chen, X.G. Superhydrophobic honeycomb-like cobalt stearate thin films on aluminum with excellent anti-corrosion properties. *Appl. Surf. Sci.* **2017**, *407*, 361–370. [CrossRef]

22. Karthik, N.; Yong, R.L.; Sethuraman, M.G. Fabrication of corrosion resistant mussel-yarn like superhydrophobic composite coating on aluminum surface. *J. Taiwan Inst. Chem. Eng.* **2017**, *77*, 302–310. [CrossRef]

23. Kang, Z.; Ye, Q.; Sang, J.; Li, Y. Fabrication of super-hydrophobic surface on copper surface by polymer plating. *J. Mater. Process. Technol.* **2009**, *209*, 4543–4547. [CrossRef]

24. Fang, W.; Wang, Y.; Li, Y.; Qian, W. Fabrication of triazinedithiol functional polymeric nanofilm by potentiostatic polymerization on aluminum surface. *Appl. Surf. Sci.* **2011**, *257*, 2423–2427.

25. Zhao, Q.; Tang, T.; Dang, P.; Zhang, Z.; Wang, F. Preparation and analysis of complex barrier layer of heterocyclic and long-chain organosilane on copper alloy surface. *Metals* **2016**, *6*, 162. [CrossRef]

26. Zhao, Q.; Tang, T.; Dang, P.; Zhang, Z.; Wang, F. The corrosion inhibition effect of triazinedithiol inhibitors for aluminum alloy in a 1 m hcl solution. *Metals* **2017**, *7*, 44. [CrossRef]

27. Mori, K.; Hirahara, H.; Oishi, Y.; Kumagai, N. Effect of triazine dithiols on the polymer plating of magnesium alloys. *Mater. Sci. Forum* **2000**, *350*, 223–234. [CrossRef]

28. She, Z.; Li, Q.; Wang, Z.; Tan, C.; Zhou, J.; Li, L. Highly anticorrosion, self-cleaning superhydrophobic Ni–Co surface fabricated on AZ91D magnesium alloy. *Surf. Coat. Technol.* **2014**, *251*, 7–14. [CrossRef]

29. Cassie, A.B.D. Wettability of porous surfaces. *Trans. Faraday Soc.* **1944**, *40*, 546–551. [CrossRef]

30. Lee, Y.; Ju, K.Y.; Lee, J.K. Stable biomimetic superhydrophobic surfaces fabricated by polymer replication method from hierarchically structured surfaces of al templates. *Langmuir* **2010**, *26*, 14103–14110. [CrossRef] [PubMed]

coatings

MDPI

Article

Effect of Surface Topography and Structural Parameters on the Lubrication Performance of a Water-Lubricated Bearing: Theoretical and Experimental Study

Zhongliang Xie [1], Zhushi Rao [2] and Huanling Liu [1,*]

[1]   School of Electro-Mechanical Engineering, Xidian University, Xi'an 710071, China; zlxie@xidian.edu.cn
[2]   Laboratory of Vibration, Shock and Noise, Shanghai Jiao Tong University, Shanghai 200240, China; xiezlsj2011@sjtu.edu.cn
*   Correspondence: hlliu@xidian.edu.cn; Tel.: +86-029-88203115

Received: 14 October 2018; Accepted: 10 December 2018; Published: 2 January 2019

Abstract: This study explored the influence of the surface topography of a bushing on the lubrication performance of a water-lubricated bearing. Bushing deformations were considered in the mathematical model. Theoretical calculations and experiments were performed. The test data corresponded well with the simulation. The main stiffness and cross stiffness coefficients were measured and compared with the theoretical values, and the empirical formula of friction coefficient was fitted based on the test data.

Keywords: water-lubricated bearing; surface topography; dynamic characteristics; empirical formula of friction coefficient; lubrication performance

---

1. Introduction

In recent decades, the rapid development of the ship-building industry has witnessed the large and extensive application of water-lubricated bearings, due to their unparalleled advantages over other bearings. Many problems have occurred in this application, which, in turn, has promoted the study of lubrication mechanism and lubrication performance of these types of bearings. A growing interest has been given to investigating the lubrication performance of water-lubricated plain journal bearings, especially the influence of the surface topography of bearing coatings.

Researchers make great efforts to improve the lubrication performance [1–3] of the bearing, such as the exploitation of surface textures [4–6], the optimum design of bearing structures [7–9], and the introduction of longitudinal grooves. Many studies [10–12] focused on the effect of surface topography [13] on the bearing performance [10,14–20]. For example, Tala-Ighil et al. [21] investigated the modeling of journal bearing characteristics. They found that the lubrication performance of the textured bearing improved significantly with appropriate surface texture geometry and texture distribution. Brito et al. [22] explored the effect of grooves in single and twin axial groove journal bearings under varying load directions. Their results also showed that the friction coefficient deceased compared with the smooth surface bearing. The characteristics of textured journal bearings with consideration of thermal effects were analyzed by Tala-Ighil et al. [23] using the finite difference method (FDM). In other references [24–29], scholars have investigated lubrication performance from different perspectives.

Dadouche et al. [30] investigated the operational performance of textured journal bearings lubricated with contaminated fluid. Special attention was focused on the load-carrying capacity, friction, and wear under different contamination levels in the lubricant. Results indicated the effectiveness of textures in capturing contaminant particles and reducing the possibility of failure. Lu et al. [31] performed experiments on the friction characteristic of journal bearings with dimpled bushings manufactured using machining and chemical etching techniques. Their results indicated that a bushing with etched dimples over the entire circumference offered a better frictional performance than a bushing with dimples etched on half of its circumference. Tala-Ighil et al. [32] presented the influence of a textured area on the lubrication performance of hydrodynamic journal bearings. Analysis indicated that a textured area on a bearing bushing could effectively increase the load-carrying capacity and increase the minimum film thickness in the main load-carrying area, while decreasing the friction coefficient at the same time. Cristea et al. [33] studied lubrication performance (transient pressure and temperature field measurements) of journal bearings with circumferential grooves, where the operating modes were lightly loaded from startup to steady-state thermal stabilization. Fluid film pressure, temperature field, friction torque, and lubricant side leakage were detected simultaneously through experimental methods. Their results indicated that film rupture starts from cavitation downstream and the minimum film thickness. This occurred because of the existence of the circumferential grooves and the surface topography. Xie et al. [34–36] explored the lubrication states, lubrication performance with consideration of the bushing macro deformation using theoretical simulation and experimental verification. Their results indicated that the surface roughness had a significant influence on the lubrication state transition and the lubrication performance. A summary of the coupled factors on the lubrication states was also given. However, even though researchers have carried out much work, investigation of the lubrication performance of plain journal bearings is rather insufficient. Only a few studies address the significance of surface topography and bushing deformation on lubrication performance, particularly the experimental verification of the bearings.

In view of the above problems, this study focused on the influence of surface topography on the lubrication performance of water-lubricated bearings. The research sheds light on the lubrication mechanism of the bearing and has a certain significance for guiding the design of such bearings.

2. Theoretical Model

2.1. Mathematic Model

Figure 1 shows the practical plain journal bearing, with the surface topography. The bearing is completely submerged in the lubricant. It carries the vertical load. Surface topography effect is considered in the analysis. For the convenience of simulation, the bushing and journal surfaces are equivalent to the bushing with combined surface roughness, while the journal is absolutely smooth (as shown in Figure 1a). Figure 1b presents the closer view of the surface roughness.

In the Cartesian coordinates, the modified Reynolds Equation with consideration of surface topography effect is as follows:

$$\frac{\partial}{\partial x}\left(\phi_x \frac{\rho h^3}{\mu}\frac{\partial p}{\partial x}\right) + \frac{\partial}{\partial z}\left(\phi_z \frac{\rho h^3}{\mu}\frac{\partial p}{\partial z}\right) = 6U\frac{\partial(\phi_c \rho h_T)}{\partial x} + 6U\sigma_s\frac{\partial(\rho\phi_s)}{\partial x} + 12\frac{\partial(\phi_c \rho h_T)}{\partial t} \tag{1}$$

Surface topography is considered using these parameters: ϕ_x, ϕ_z are the pressure flow factors, ϕ_s is the shear flow factor, and ϕ_c is the contact factor.

For the Equation above, the pressure distribution of the fluid field is governed by the structure and operating parameters, as well as the film thickness.

The modified formula of the film thickness is as follows:

$$h = h_0 + \delta h + \delta_1 + \delta_2 \tag{2}$$

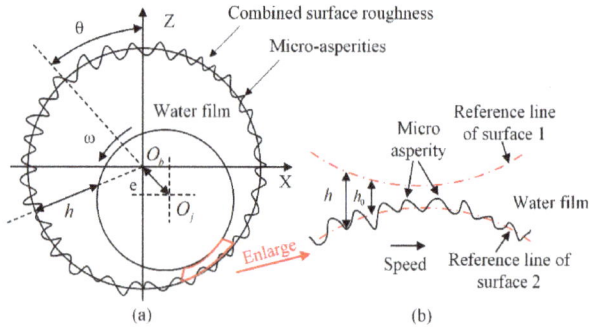

Figure 1. Schematic representation of the film thickness in contact with rough surfaces.

δh is the macroscopic elastic deformation of the bearing bushing. The macroscopic elastic deformation of the bushing surface due to the normal hydrodynamic effect is calculated using the following Boussinesq Formula:

$$\delta h = \frac{2}{\pi E'} \iint_{\Omega} \frac{p(\xi, \zeta)}{\sqrt{(x - \xi)^2 + (y - \zeta)^2}} d\xi d\zeta \tag{3}$$

If a groove is considered in the model, the film thickness should be modified as follows:

$$h = \begin{cases} h_0 + \delta h + \delta_1 + \delta_2, & 0 \leq \theta \leq \theta_s, \theta_f \leq \theta \leq 2\pi \\ h_0 + \delta h + \delta_1 + \delta_2 + \delta_{groove}, & \theta_s < \theta_f < \theta \end{cases} \tag{4}$$

where θ_s, θ_f are the start angle and the end angle of the groove, respectively δ_{groove} is the height of the fluid film in the groove.

$$\delta_{groove} = d_0 - C(1 + \cos(\theta - \pi/2))$$

The surface plastic macro deformation can also be added into the calculation if needed. However, for composite materials due to their wear resistance, impact, and corrosion properties, the calculation of plastic macro deformation is rather cumbersome. The equilibrium Equation for the hydrodynamic force, contact force, and external load $\overrightarrow{P}$ are calculated by:

$$\overrightarrow{P} + \overrightarrow{F}_{fluid} + \overrightarrow{W}_{asp} = 0 \tag{5}$$

where hydrodynamic force $\overrightarrow{F}_{fluid} = \int_0^1 \int_0^{2\pi} pd\theta dz$ and film pressure are determined by the mixed lubrication (ML) model. Contact force $\overrightarrow{W}_{asp} = \int_0^1 \int_0^{2\pi} p_{asp} d\theta dz$.

Formula of friction coefficient:

$$\begin{aligned} f_{total} &= \alpha_1 f_{asp} + \alpha_2 f_{fluid} \\ \alpha_1 + \alpha_2 &= 1, \ \alpha_1, \alpha_2 \in [0, 1] \end{aligned} \tag{6}$$

where f_{asp} is the friction coefficient due to micro-asperities contacts effect and f_{fluid} is the friction coefficient due to "viscous effect" (i.e., the shearing stress of the fluid molecules).

2.2. Boundary Conditions

For water-lubricated bearings, the commonly used boundary conditions include the Reynolds, Jakobsson-Floberg–Olsson (JFO), and Sommerfeld boundary conditions. For different boundary conditions, the calculated lubrication performance differs.

For JFO boundary conditions [20], the calculated pressure distribution is limited to the positive pressure region, and both the cavitation upper and lower boundaries are based on the mass conservation equation. In the cavitation region, the pressure is equal to the cavitation pressure P_{cav}, whereas in the rupture region (x_0, z_0),

$$P_{x_0, z_0} = P_{cav}, \quad \left(\frac{\partial P}{\partial x}\right)\bigg|_{x=x_0} = \left(\frac{\partial P}{\partial z}\right)\bigg|_{z=z_0} = 0 \tag{7}$$

For Reynolds boundary conditions, the nature rupture boundary condition, which considers the water film to be continuous, and the end point of the film are a nature rupture phenomenon. The film will fracture automatically after striking the minimum film thickness. Generally speaking, Reynolds boundary conditions are closer to the practical engineering operations:

$$\begin{cases} z = 0, P = P_a, z = z_2, \frac{dP}{dz} = 0 \\ 0 < z < z_2, P = P(z), z_2 < z < 2\pi, P = P_a, \frac{dP}{dz} = 0 \end{cases} \tag{8}$$

In this study, the above two boundary conditions are used to investigate the influence on lubrication performance.

For the bearing-rotor system, when the rotor is disturbed, the support force of the film will change correspondingly. If the disturbance is small, the force can be expanded through the Taylor series near the equilibrium point, then:

$$\begin{cases} F_x = F_{x_0} + \frac{\partial F_x}{\partial x}\Delta x + \frac{\partial F_x}{\partial y}\Delta y + \frac{\partial F_x}{\partial x}\Delta \dot{x} + \frac{\partial F_x}{\partial y}\Delta \dot{y} + \cdots \\ F_y = F_{y_0} + \frac{\partial F_y}{\partial x}\Delta x + \frac{\partial F_y}{\partial y}\Delta y + \frac{\partial F_y}{\partial x}\Delta \dot{x} + \frac{\partial F_y}{\partial y}\Delta \dot{y} + \cdots \end{cases} \tag{9}$$

where F_x, F_y are the components of the film forces and F_{x0}, F_{y0} are the components of the film forces on the equilibrium position. When the disturbance is small, the high order components of the second order and above are ignored, whereas the first order components are retained:

$$\begin{cases} \Delta F_x = F_x - F_{x_0} = k_{xx}\Delta x + k_{xy}\Delta y + c_{xx}\Delta \dot{x} + c_{xy}\Delta \dot{y} \\ \Delta F_y = F_y - F_{y_0} = k_{yx}\Delta x + k_{yy}\Delta y + c_{yx}\Delta \dot{x} + c_{yy}\Delta \dot{y} \end{cases} \tag{10}$$

For the dynamic coefficients of the film:

$$\begin{cases} k_{xx} = \frac{\partial F_x}{\partial x}\big|_0, k_{xy} = \frac{\partial F_x}{\partial y}\big|_0, k_{yx} = \frac{\partial F_y}{\partial x}\big|_0, k_{yy} = \frac{\partial F_y}{\partial y}\big|_0 \\ c_{xx} = \frac{\partial F_x}{\partial x}\big|_0, c_{xy} = \frac{\partial F_x}{\partial y}\big|_0, c_{yx} = \frac{\partial F_y}{\partial x}\big|_0, c_{yy} = \frac{\partial F_y}{\partial y}\big|_0 \end{cases} \tag{11}$$

$$\begin{pmatrix} k_{xx} \\ k_{yx} \end{pmatrix} = \int_0^1 \int_0^{2\pi} p_x \begin{pmatrix} \sin x \\ -\cos x \end{pmatrix} dx dz \tag{12}$$

$$\begin{pmatrix} k_{xy} \\ k_{yy} \end{pmatrix} = \int_0^1 \int_0^{2\pi} p_y \begin{pmatrix} \sin x \\ -\cos x \end{pmatrix} dx dz \tag{13}$$

$$\begin{pmatrix} c_{xx} \\ c_{yx} \end{pmatrix} = \int_0^1 \int_0^{2\pi} p_x \begin{pmatrix} \sin x \\ -\cos x \end{pmatrix} dx dz \tag{14}$$

$$\begin{pmatrix} c_{xy} \\ c_{yy} \end{pmatrix} = \int_0^1 \int_0^{2\pi} p_y \begin{pmatrix} \sin x \\ -\cos x \end{pmatrix} dx dz \tag{15}$$

3. Experimental Section

3.1. Test Apparatus

Figure 2 presents the experimental apparatus, and Figure 3 illustrates the simple line drawing of the system. For this study, a multi-function bearing-rotor coupled system was used. Different types of bearings could be tested, for example, plain journal bearings, rolling element bearings, and thrust bearings. The shaft was supported by two hydrostatic bearings between the coupling and the test bearing. These two hydrostatic bearings stiffened the rotor system and decreased the deformation of the shaft when vertical load was applied on the tested bearing. They decreased the influence of overhung load and improved the accuracy of the system.

The lubricant tank contained the lubricant (for this test, it was water), and the test bearing was completely immersed into the lubricant to guarantee full film lubrication. For the lubricant, water at room temperature (20 °C) was used. Then, the bearing was fully submerged into the water tank; thermal effects were negligible during the experiment. In the case of water, the viscosity is almost independent of the temperature, which means that the viscosity remains constant.

Four displacement sensors were installed in the housing along the circumferential direction. The test bearing was subjected to external load through a hydraulic cylinder system. Maximum velocity of the shaft was 6000 rpm. Specific pressure for the bearing was 0–2 MPa, which could be adjusted according to the practical operating mode.

The test bearing was subjected to external load through the vertical load unit. Figure 4a presents the load unit on the bearing. Two force sensors and four displacement sensors were installed on the apparatus. Force sensor #1 measured the external load in the vertical direction, and Force sensor #2 measured the tangential force in the circumferential direction. Four displacement sensors were uniformly distributed along the circumferential direction and measured the rotor vibration and the film thickness. The bearing was completely immersed into the lubricant in order to guarantee full-film lubrication. Figure 4b gives a closer view of the external load unit.

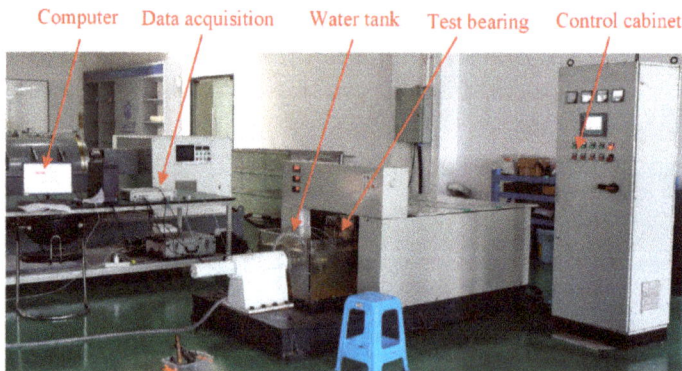

Figure 2. Physical layout of the bearing-rotor test rig system.

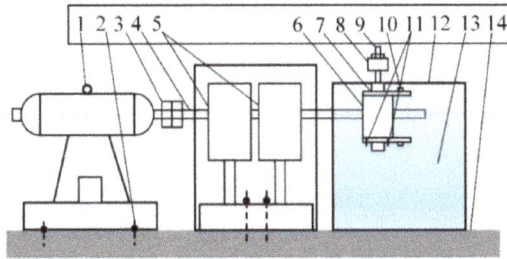

Figure 3. Line drawing of the experiment system: 1-motor; 2-damper; 3-coupling; 4-shaft; 5-hydrostatic bearing; 6-tested bearing; 7-shaft sleeve; 8-hydraulic cylinder; 9-leading bar; 10-force sensor; 11-eddy–current sensor; 12-water tank; 13-water; 14-base.

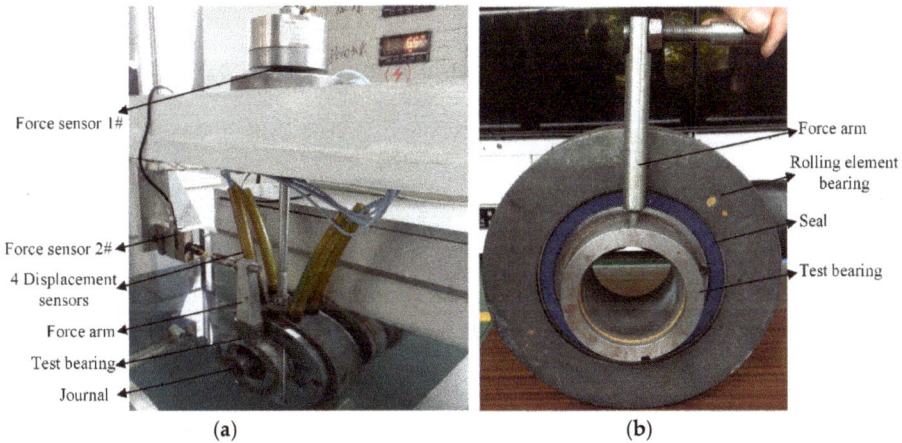

Figure 4. Closer view of the load unit. (**a**) load unit; (**b**) closer view of the load unit.

3.2. Experimental Procedure

In order to measure the film thickness, four displacement sensors were mounted along the circumferential direction (as can be seen in Figure 5a,b). If the exact value of the radial clearance is C_0, then assume the initial value of the radial clearance to be C. With the given parameters, the film thickness distribution h and the pressure distribution p can be calculated theoretically. Furthermore, the distances between the bushing and shaft can also be obtained: $d_{10}, d_{20}, d_{30}, d_{40}$. At the same time, the distances can also be measured: d_1, d_2, d_3, d_4.

If $|d_{10} - d_1| \leq 0.5\mu m, |d_{20} - d_2| \leq 0.5\mu m, |d_{30} - d_3| \leq 0.5\mu m, |d_{40} - d_4| \leq 0.5\mu m$, the radial clearance of the bearing is C. Otherwise, one can modify the value of the clearance and restart the loop until it is convergent. The convergent criterion was 0.5 μm. Then, after the minimum value of the film thickness distribution is found, the minimum film thickness h_{min} is obtained.

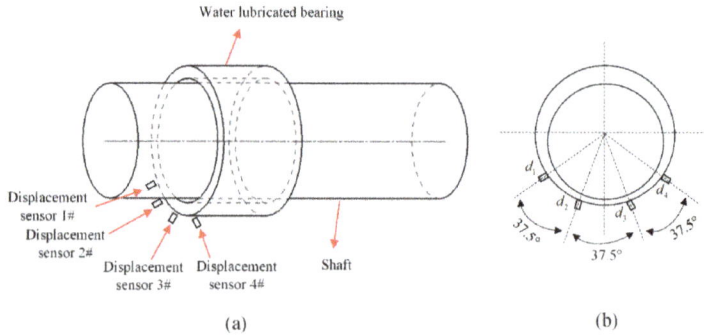

Figure 5. Layout of measuring points for the bearing. (**a**) displacement sensors along the circumferential; (**b**) relative positions of the displacement sensors.

3.3. Test Bearing

Geometry parameters of the test bearings can be found in Table 1. The measured dry friction coefficient between steel and bushing is 0.13. Therefore, in order to explore the effect of surface topography on lubrication performance of the bearing, several sets of test bearings with different surface roughness Sa were processed (as can be seen in Table 2).

Table 1. Geometry of the bearing.

Description	Symbol	Value	Dimension
Width	L	80	mm
Diameter	D	62	mm
L/D ratio	L/D	1.30	–
Radial clearance	C	0.03–0.07	mm
Clearance ratio	Ψ	0.096‰–2.25‰	–
Velocity	V	0.001–10	m/s
External load	F	80–6000	N

In the experiment, the following parameters were measured: F is the total external load exerted on the bearing, which includes the bearing gravity, T_f is the tangential force, and $T_f \times R$ is the friction torque. Other parameters include the following: D is the diameter and G is the bearing gravity.

The force applied on the bearing is as follows:

$$F_r = F - G \tag{16}$$

The coefficient of friction of the bearing is as follows:

$$f = \frac{2 \times T_f \times R}{F_r \times D} \tag{17}$$

For the test bearing, $R = 205$ mm, $D = 62$ mm, $G = 150$ N, the expression of the friction coefficient is as follows:

$$f = \frac{205 T_f}{31 F_r} \tag{18}$$

The measuring accuracy of the eddy–current sensor was 0.1 μm, whereas the sampling frequency was 1000 Hz. The nonlinear error $\leq \pm 0.1\%$, and the response frequency was 10 kHz. The measuring accuracy of the force sensor was 0.1 N. During the test, the data acquisition system should balance to ensure the equilibrium position of the shaft. The test data were recorded about 180 s under each working mode. Before the data acquisition, the test bearing should run for several minutes. The friction values fluctuate periodically with time with a small amplitude. A more detailed description of the test bearing can be found in the reference [34,35].

4. Results and Discussion

4.1. Measurement of Surface Topography

Figure 6a–d shows the 3D morphology distribution of the four test bearings. Table 2 gives the characteristic parameters of the surface topography for the four tested samples. S_a is the arithmetic mean height; S_q is the root mean square height; V_{mp} is the peak material volume, which represents the part that will be worn out in the test; and V_{vv} is the pit void volume.

One thing to note for scientific validity is that more than one sample extract per surface should be investigated. In the experiment, dozens of samples were extracted for one surface. Characteristic surface parameters were measured for the samples. Mathematic algorithms exist within the Universal Profilometer that can automatically deduce the characteristic surface parameters for the surface. In this study, we only present one 3D morphology distribution contour for each test surface.

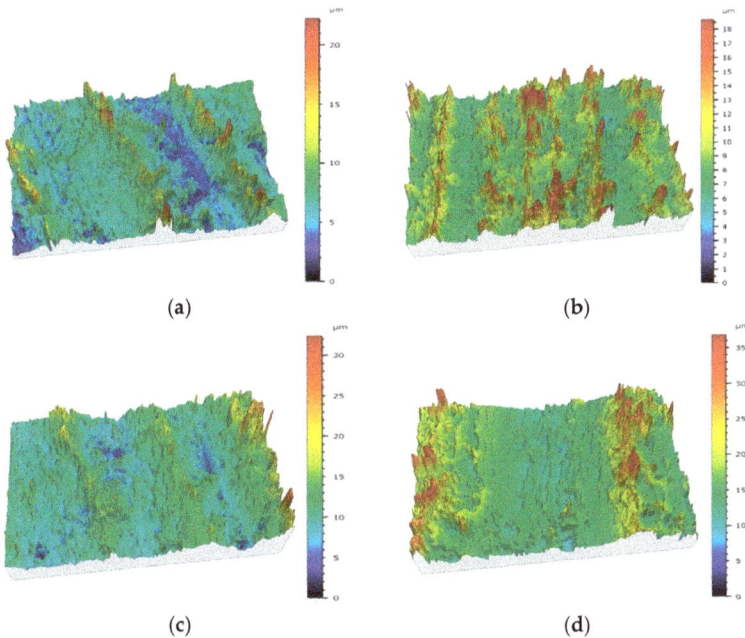

Figure 6. 3D morphology distribution of the four test samples (the height scale is not the same in all subfigures). (**a**) sample 1#; (**b**) sample 2#; (**c**) sample 3#; (**d**) sample 4#.

From Table 2 we can see the variation of the real surface topography for the four tested samples. For samples #1, #2, #3, and #4, the arithmetic mean height and the root mean square height increase correspondingly with the surface roughness. After the experiment, characteristic parameters were significantly reduced, which shows the improvement of the surface topography.

Typically, in the start-up and shut-down stage of the bearing-rotor coupled system, the water-lubricated plain journal bearing is at low speed and heavy load state and the film is thin (film thickness may be less than 10 μm in most cases). In some extreme operating conditions, the film thickness may be just a few microns. Water film thickness has almost the same order of magnitude as the bushing interface topography (machining accuracy of composite material bushing >1.6 μm), thus the film thickness ratio is very small. Direct contacts of micro-asperities take place under certain conditions.

These undoubtedly affect the fluid hydrodynamic. Therefore, we will investigate further and experimentally verify the lubrication state transitions of the bearing.

Table 2. Characteristic parameters of the surface topography for the four tested samples. S_a: arithmetic mean height; S_q: root mean square height; V_{mp}: peak material volume; V_{vv}: pit void volume.

Parameter	Before/After Experiment	Unit	S1	S2	S3	S4
S_a	Before	µm	1.439 (±0.0288)	2.046 (±0.0409)	2.107 (±0.0421)	3.134 (±0.0627)
	After	µm	1.222 (±0.0244)	1.914 (±0.0383)	1.853 (±0.0371)	2.238 (±0.0448)
S_q	Before	µm	1.995 (±0.0399)	2.656 (±0.0531)	2.821 (±0.0564)	4.028 (±0.0842)
	After	µm	1.668 (±0.0334)	2.216 (±0.0452)	2.029 (±0.0406)	2.870 (±0.0574)
V_{mp}	Before	µm³/µm²	0.183 (±0.0037)	0.197 (±0.0039)	0.230 (±0.0046)	0.323 (±0.0065)
	After	µm³/µm²	0.141 (±0.0028)	0.158 (±0.0032)	0.208 (±0.0041)	0.305 (±0.0061)
V_{vv}	Before	µm³/µm²	0.171 (±0.0034)	0.206 (±0.0041)	0.254 (±0.0051)	0.259 (±0.0052)
	After	µm³/µm²	0.165 (±0.0033)	0.162 (±0.0032)	0.211 (±0.0042)	0.213 (±0.0043)

4.2. Verification of the Model

The accuracy and reliability of the model and algorithm were prerequisites for the study. The accuracy of the model was verified by comparing the results with reference [20]. Figure 7 presents the schematic diagram of the relative position of Sections 1 and 2 on the test bearing. Sections 1 and 2 are the two separate sections on the bearing in the axial direction. Section 1 is located at the 1/5 L of the bearing, whereas Section 2 is located at the 3/5 L of the bearing. Figure 8 shows the comparison of pressure distribution between this research and the reference, for Sections 1 and 2 (as shown in Figure 7). Pressure data in the circumferential direction along the liquid film at the two sections of the bearing were extracted and compared with the theoretical analysis and experimental results in the reference, as shown in Figure 8.

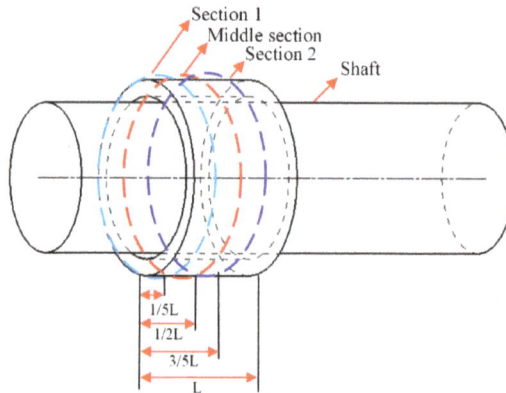

Figure 7. Schematic diagram of Sections 1 and 2.

From Figure 8, we can see that this study's simulation corresponds well with the literature under two boundary conditions, the Reynolds boundary condition and the Jakobsson-Floberg-Olsson (JFO) boundary condition, which proves the accuracy of the model and algorithm. Pressure under the Reynolds boundary conditions shows better correspondence with the reference [20]. The following calculation is based on this model under the Reynolds boundary conditions.

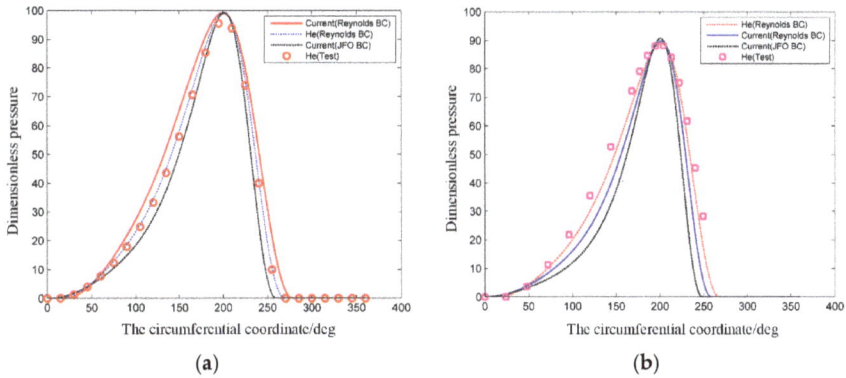

Figure 8. Comparison of the pressure distribution between this research and reference [20]. (**a**) Section 1; (**b**) Section 2.

4.3. Effect of Elastic Deformation of the Bushing

Figure 9 shows the effect of eccentricity ratio epsilon on the pressure distribution in the circumferential direction. As the eccentricity ratio epsilon increases, the maximum value of pressure increases, which shows a strong nonlinearity. At the same time, the main load-carrying area decreases.

Figure 10 shows the effect of bushing thickness on the pressure distribution in the circumferential direction. We observe that the thicker the bushing, the smaller the film maximum pressure. The main load-carrying area also increases with the bushing thickness increase. Correspondingly, the load-carrying capacity and friction coefficient decrease. This can be explained from the perspective of energy, that is, if the bearing bushing is absolutely rigid without elastic deformation, the external load will be balanced by the film thickness. However, under the same conditions, the bushing deformation will absorb part of the energy and share the external load together with the water film. The thicker the bushing, the more energy the bushing will absorb thereby reducing the maximum pressure of the film. This is also the bushing's mechanism of cushion and shock absorption.

Figure 11 shows the effect of elastic deformation on the dimensionless load-carrying capacity and the friction force. The rigid model represents the model which does not consider the deformation of the bushing, whereas the elastic model considers the deformation. Load-carrying capacity and friction force both increase with the eccentricity ratio. Load-carrying capacity with elastic deformation is smaller than that of the rigid model, whereas friction force is higher than that of the rigid model. The difference between the rigid and elastic models gradually increases with the eccentricity ratio.

Figure 12 shows the effect of elastic deformation on the dynamic coefficients. With the increase in the eccentricity ratio, the dynamic coefficients K_{xx}, K_{xy}, K_{yx}, K_{yy}, C_{xx}, C_{xy}, C_{yx}, C_{yy}, increase for the rigid and elastic models. Dynamic coefficients with elastic deformation are smaller than that of the rigid model. The existence of elastic deformation decreases the rigidity of the bearing, which is beneficial for the stability of the bearing-rotor system. Elastic deformation decreases the maximum pressure and dynamic characteristics. The difference between the rigid and elastic models gradually increases with the eccentricity ratio.

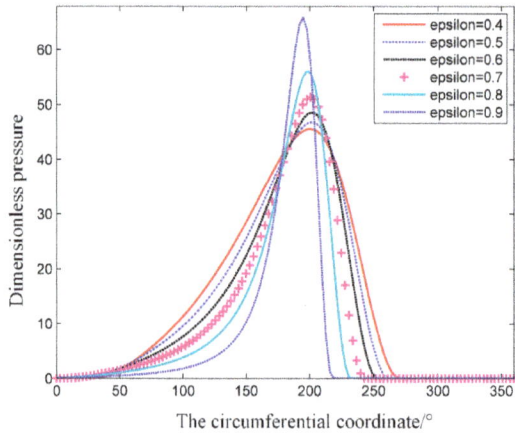

Figure 9. Effect of eccentricity ratio on the pressure distribution.

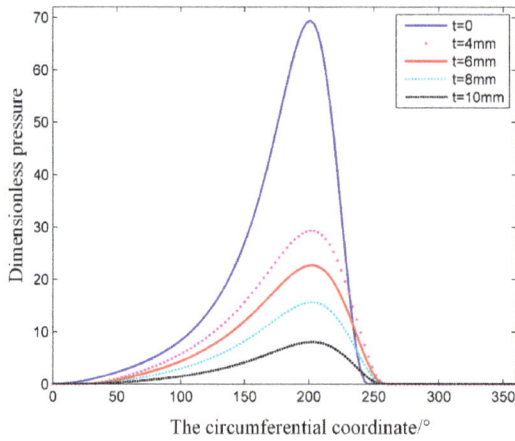

Figure 10. Effect of bushing thickness on the pressure distribution.

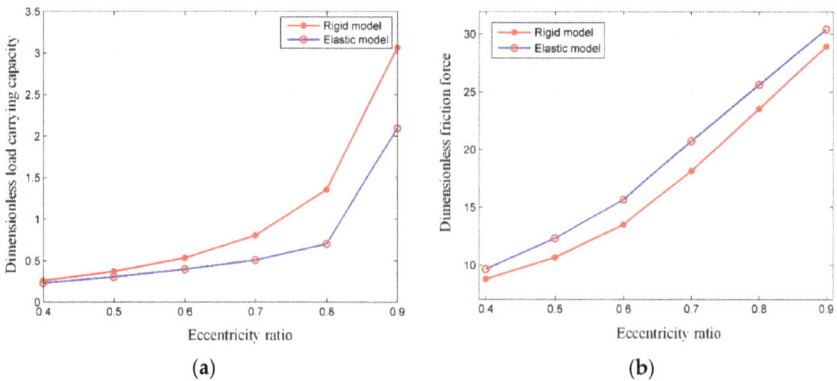

Figure 11. Effect of elastic deformation on the lubrication performance: (**a**) load-carrying capacity; (**b**) friction coefficient.

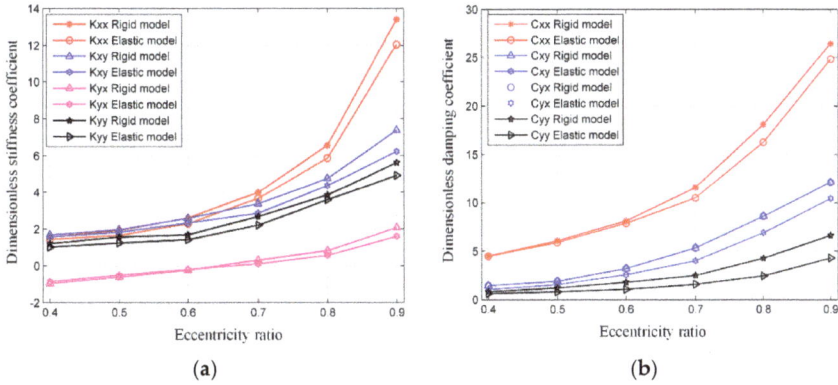

Figure 12. Effect of elastic deformation on the dynamic coefficients: (**a**) stiffness coefficient; (**b**) damping coefficient.

4.4. Minimum Film Thickness

The minimum film thickness is the major judging criterion for the lubrication regime's transition and the lubrication performance. Specific pressure, velocity and viscosity are the dominant influencing factors on the minimum film thickness. It is of vital significance to examine the minimum film thickness and its influencing factors.

Figure 13 presents the measured minimum film thickness as a function of velocity under different specific pressures for the test bearing. Under the same specific pressure, the minimum film thickness increases with the velocity. For different specific pressures, the changing rule is different. When the bearing is subjected to a light load (specific pressure of 0.125 MPa), the minimum film thickness rises sharply in the low velocity range, rises moderately in the medium velocity range, and remains the same in the high velocity range, which illustrates that the contribution of velocity to minimum film thickness is greater than that of specific pressure. When the bearing is subjected to a medium load (specific pressure of 0.156, 0.313 and 0.625 MPa), the minimum film thickness increases almost linearly over the entire velocity range. In the case of a high load condition (specific pressure of 0.938 MPa), the minimum film thickness increase slowly with the velocity over the entire velocity range, which indicates that the contribution of specific pressure to minimum film thickness is greater than that of velocity.

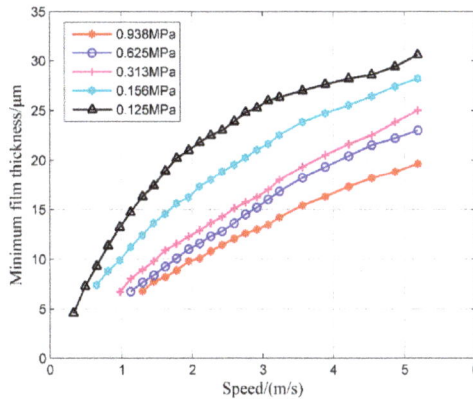

Figure 13. Minimum film thickness under different specific pressures.

4.5. Stiffness Coefficients

The stiffness of the bearing in the vertical direction K_{yy} is equal to the ratio of the load increment ΔF_y in the y direction to the displacement increment Δy in the y direction. The expression is as follows:

$$K_{yy} = \frac{\Delta F_y}{\Delta y} = \frac{F_{y2} - F_{y1}}{y_2 - y_1} \tag{19}$$

Figure 14 shows the relationship between the external load and the relative displacement in the y direction. When the load is in the range 1.50–4.50×10^3 N, the linearity of the curve is better. The slope of the curve represents the stiffness coefficient, $K_{yy} = 2.1 \times 10^7$ N m^{-1}, and its theoretical value is $K_{yy} = 2.86 \times 10^7$ N m^{-1} with a relative error of 26.57%, as shown in Table 3.

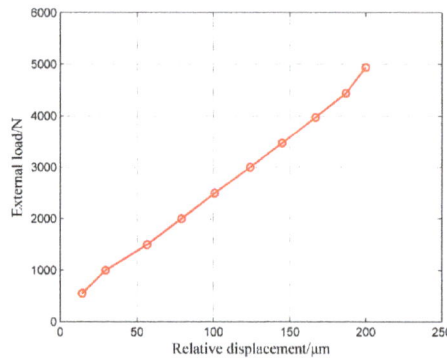

Figure 14. Curve of the external load and the relative displacement in the vertical direction.

Similarly, the cross stiffness coefficient K_{yx} is equal to the ratio of the load increment ΔF_y in the y direction to the displacement increment Δx in the x direction. The expression is as follows:

$$K_{yx} = \frac{\Delta F_y}{\Delta x} = \frac{F_{y2} - F_{y1}}{x_2 - x_1} \tag{20}$$

For the same conditions as in Figure 14, Figure 15 presents the relationship between the external load and the relative displacement in the x direction. The slope of the curve represents the stiffness coefficient, $K_{yx} = 2.72 \times 10^6$ N m^{-1}, and its theoretical value is $K_{yx} = 3.29 \times 10^6$ N m^{-1} with a relative error of 21.00%, as shown in Table 3.

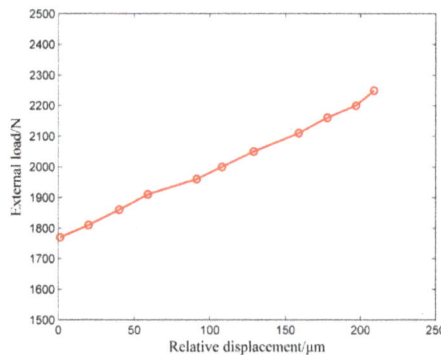

Figure 15. Curve of the external load and the relative displacement in the horizontal direction.

Table 3. Dynamic coefficient of the bearing.

Item	Theoretical (N/m)	Test (N/m)	Relative Error
K_{yy}	2.86×10^7	2.10×10^7	26.57%
K_{yx}	3.29×10^6	2.72×10^6	21.00%

4.6. Empirical Formula of Friction Coefficient

Figure 16 shows the experimental verification of the friction coefficient for two test bearings with different surface roughness. Specifically, σ_1 and σ_2 represent bearings #1 and #2, respectively. Test data correspond well with the simulation, for the high velocity region, and the relative error is very small, especially for bearing #2.

For the practical engineering bearings, it is generally difficult to accurately measure the film thickness. Friction coefficient is the commonly used parameter, which can be measured exactly [37–40].

Figure 17a,b shows the relationship between the friction coefficient and the velocity under different clearance ratios, namely, (a) $\psi = 0.5\%$, (b) $\psi = 1.0\%$, (c) $\psi = 1.5\%$, and (d) $\psi = 2.0\%$. From the figures we observe that as the velocity increases, the friction coefficient decreases sharply in the low speed range, and decreases moderately in the medium and high speed range.

Figure 18a,b shows the relationship between the friction coefficient and the specific pressure under different clearance ratios, namely, (a) $\psi = 0.5\%$, (b) $\psi = 1.0\%$, (c) $\psi = 1.5\%$, and (d) $\psi = 2.0\%$. From the figures we observe that the friction coefficient decreases moderately with the increase in the specific pressure over the entire speed range [41].

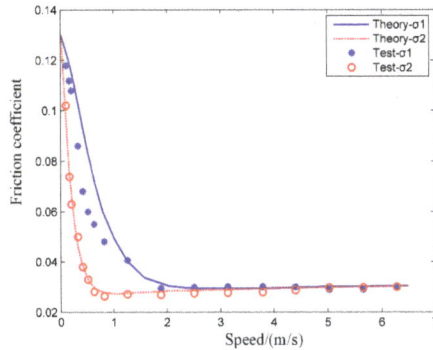

Figure 16. Verification of friction coefficient (σ_1 represents bearing #1, σ_2 represents bearing #2).

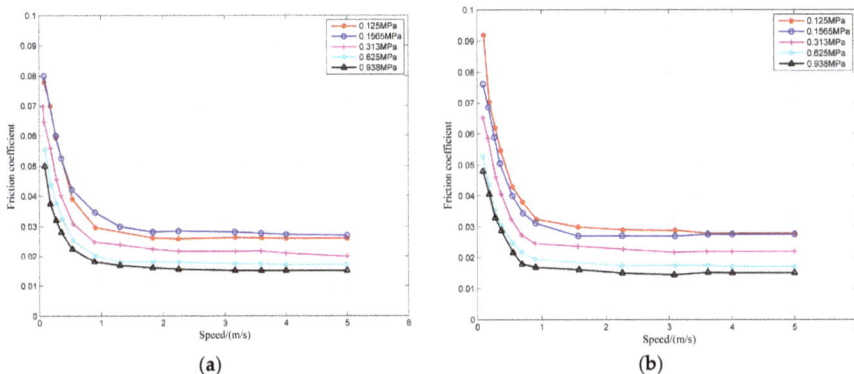

(a)

(b)

Figure 17. *Cont.*

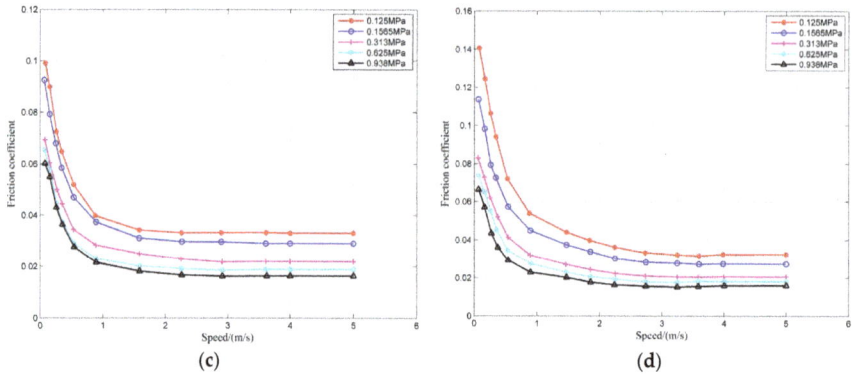

Figure 17. Relationship between the friction coefficient and the velocity for different clearance ratios ψ: (a) ψ = 0.5‰, (b) ψ = 1.0‰, (c) ψ = 1.5‰, and (d) ψ = 2.0‰.

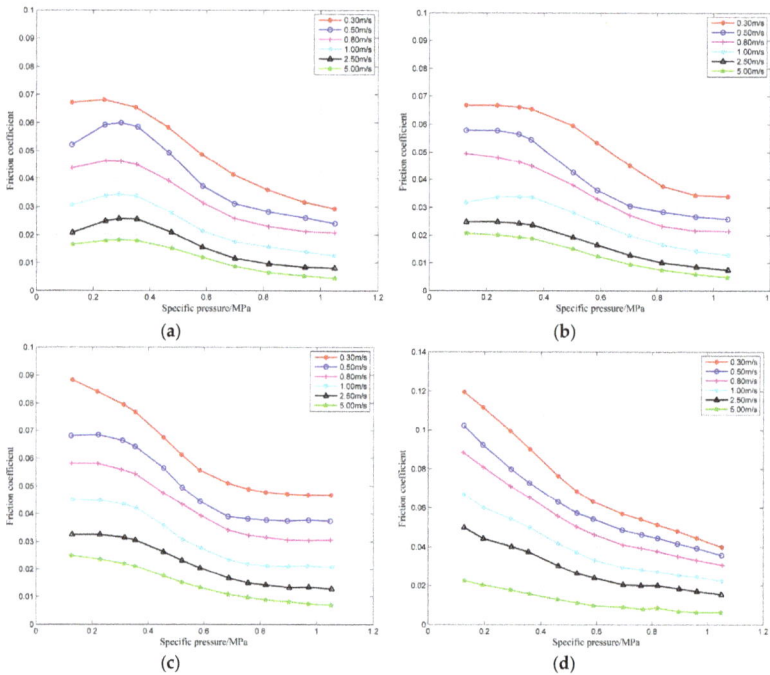

Figure 18. Relationship between the friction coefficient and the specific pressure for different clearance ratios ψ: (a) ψ = 0.5‰, (b) ψ = 1.0‰, (c) ψ = 1.5‰, and (d) ψ = 2.0‰.

Generally speaking, it is of great concern to estimate the friction coefficient and evaluate the lubrication performance. The main influencing factors affecting the bearing's performance include the specific pressure, the clearance ratio, the velocity, and the width to diameter ratio. The friction coefficient is closely related to these parameters. Based on the tested data, the relationship between the friction coefficient and the above factors was obtained through fitting. The effects of the bearing's structural parameters and operating conditions on tribological properties were examined. The empirical formula developed with velocity, specific pressure, and clearance ratio is as follows:

$$f(v, p, \psi) = \alpha_1 \cdot e^{(-\alpha_2 v + \alpha_3)} \cdot p^{\alpha_4} \cdot \psi^{\alpha_5} + \alpha_6 \tag{21}$$

where p is the specific pressure, which is equal to the external load divided by the equivalent loaded area, $p = \frac{F_{total}}{A}$, with the unit N m^{-2}; A is the equivalent loaded area, which is equal to the diameter multiplied by the width, $A = D \times L$, with the unit m^2; ψ is the clearance ratio, which is equal to the radial clearance divided by the eccentricity, $\psi = \frac{c}{e}$; α_1 to α_6 are coefficients and through fitting experimental data, estimated values can be obtained. Table 4 gives the estimated values of the coefficients α_1 to α_6. The range of values for α_1 to α_6 and the recommended values for α_1 to α_6 are also found in Table 4.

Table 4. Estimated values of the coefficients.

Item	Range of the Value	Recommended Value
α_1	0.02468–0.1206	0.0967
α_2	5.5180–7.8470	6.6460
α_3	0.1018–0.2226	0.1531
α_4	0.1035–0.4069	0.2745
α_5	0.3372–0.6800	0.4439
α_6	0.02758–0.02997	0.0288

Thus, the empirical formula of the friction coefficient obtained through fitting the measured data is as follows:

$$f_1(v, p, \psi) = 0.0967 \cdot e^{(-6.646v+0.1531)} \cdot p^{0.2745} \cdot \psi^{0.4439} + 0.0288 \tag{22}$$

Theoretical calculations, experimental results and the fitting data are plotted in Figure 19, where in (a) the clearance ratio is 1.0‰ and in (b) the clearance ratio is 1.5‰. Fitting curves correspond well with the experimental data, while some deviation occurs compared to the theoretical calculations. Nevertheless, the empirical formula is beneficial for the optimum design of structures, the rationale for selecting the parameters (e.g., the clearance ratio, the specific pressure) and the prediction of tribological characteristics.

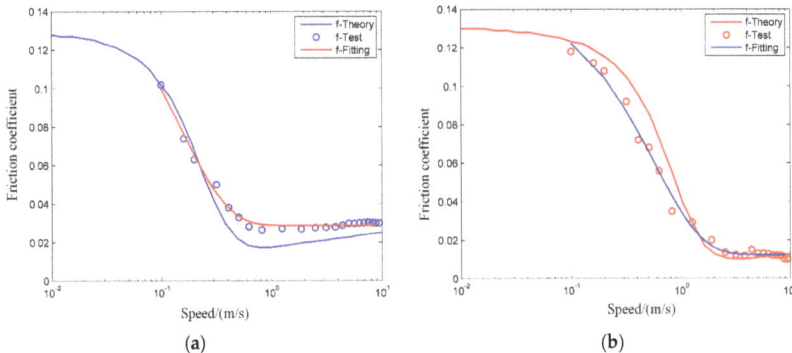

Figure 19. Fitting curve of the friction coefficient: (a) ψ=1‰, (b) ψ=1.5‰.

The empirical formula can be used to predict the friction coefficient of the bearing. Figure 20 shows the comparison of the empirical formula values and the experimental data. For P_1 and P_2 the specific pressure is 0.15 and 0.20 MPa, respectively, and the clearance ratio is 1.0‰. Predicted values using the empirical formula correspond well with experimental data in the medium and high velocity region, while deviations occur in the low velocity region.

Figure 21 shows the three-dimensional distribution of the friction coefficient with the velocity and the clearance ratio. From Figure 21 it is clear that the optimum range of the clearance ratio is approximately from 0.8‰ to 2.0‰. This is of important significance for guiding the parameter optimization and structure design of such bearings.

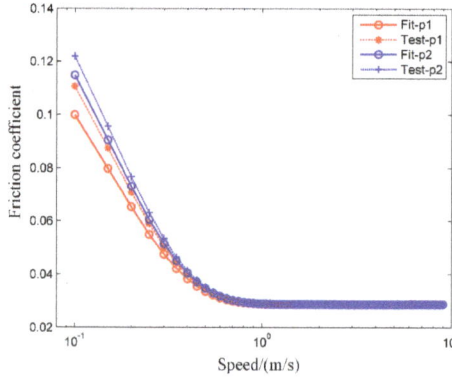

Figure 20. Fitting curve of the friction coefficient ($\psi = 1‰$).

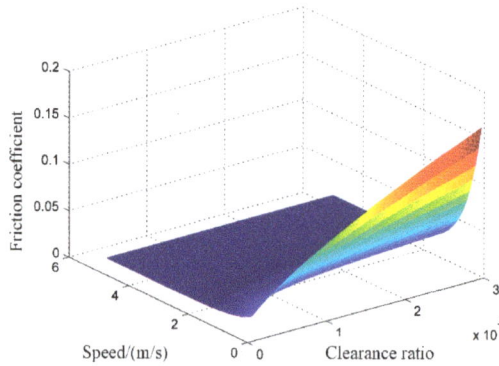

Figure 21. Fitting curve of the friction coefficient.

5. Conclusions

This study has a certain significance for guiding the future investigation of the lubrication state transition of water-lubricated plain journal bearings. An empirical formula was proposed using test data fitting. Predicted values corresponded well with the experiment, and they will be beneficial for the optimum structure design of the bearing. From the analysis, the following conclusions can be made:

- The existence of bushing decreases the dimensionless pressure. With the increase in the bushing thickness, the dimensionless pressure decreases correspondingly;
- With the increase in the eccentricity ratio, the dimensionless load-carrying capacity and the friction force increase. The existence of bushing deformation (elastic model) decreases the load-carrying capacity but increases the friction force;
- With the increase in the eccentricity ratio, the dimensionless stiffness and damping coefficients increase. The existence of the bushing deformation (elastic model) decreases the dynamic characteristic coefficients;
- Under the same specific pressure, with the increase in the speed, the minimum film thickness increases. Under the same speed, with the increase in the specific pressure, the minimum film thickness decreases. Specific pressure and velocity are the dominant influencing factors on the measured minimum film thickness;

- The empirical formula of friction coefficient with velocity, specific pressure, and clearance ratio is obtained based on experimental data. The empirical formula is beneficial for the optimum design of structures and the prediction of tribological characteristics.

These conclusions are useful for the structure design, analysis, and optimization of journal bearings. For future research, we will continue to improve the test rig and the measurement accuracy. The mathematical model will also need to consider more comprehensive influencing factors.

Author Contributions: Conceptualization, Z.X. and Z.R.; Data curation, Z.X.; Formal analysis, Z.X.; Funding acquisition, Z.X. and Z.R.; Investigation, Z.X. and H.L.; Methodology, Z.X.; Project Administration, Z.X. and Z.R.; Resources, Z.X.; Software, Z.X.; Supervision, Z.X.; Validation, Z.X.; Visualization, Z.X.; Writing–Original Draft, Z.X.; Writing–Review and Editing, Z.X., Z.R., and H.L.

Funding: This research was funded by the National Natural Science Foundation of China (No. 1167020010).

Acknowledgments: The authors thank Chunxiao Jiao and Xiuli Zhang in Shanghai Jiao Tong University for their contributions to the experiments.

Conflicts of Interest: The authors declare no conflict of interest.

Nomenclature

C	radial clearance = $R_b - R_j$
R_b, R_j	bearing and journal radii
e	eccentricity
ε	eccentricity ratio, e/c
L, D	width and diameter of bearing
p	hydrodynamic pressure
h_{min}	minimum film thickness
δh	macroscopic elastic deformation of the bushing
h	real film thickness
h_0	nominal film thickness
$\overrightarrow{P}$	total external load
ω	angular velocity = $2\pi N$
ρ	lubricant density
μ	lubricant viscosity
T_f	tangential force
$k_{xx}, k_{xy}, k_{yx}, k_{yy}$	coefficient of stiffness
O_b, O_j	bearing, journal left
θ	angular coordinate
σ_1, σ_2	RMS surface roughness of two surfaces
δ_1, δ_2	roughness height of two surfaces
σ	combined surface roughness
λ	film thickness ratio
T	bushing thickness
E	combined elastic modulus
ϕ_s	shear flow factor
ϕ_c	contact factor
ϕ_x, ϕ_z	pressure flow factors
θ_s, θ_f	start angle and end angle of the groove
δ_{groove}	height of fluid film in the groove
ψ	clearance ratio
G	bearing gravity
$c_{xx}, c_{xy}, c_{yx}, c_{yy}$	coefficient of damping

References

1. Hirani, H.; Verma, M. Tribological study of elastomeric bearings for marine propeller shaft system. *Tribol. Int.* **2009**, *42*, 378–390. [CrossRef]
2. Huang, W.; Xu, Y.; Zheng, Y.; Wang, X. The tribological performance of Ti(C,N)-based cermet sliding against Si_3N_4 in water. *Wear* **2011**, *270*, 682–687. [CrossRef]
3. Wang, J.; Yan, F.; Xue, Q. Tribological behavior of PTFE sliding against steel in sea water. *Wear* **2009**, *267*, 1634–1641. [CrossRef]
4. Wang, N.; Meng, Q.; Wang, P.; Geng, T.; Yuan, X. Experimental Research on Film Pressure Distribution of Water-Lubricated Rubber Bearing with Multiaxial Grooves. *J. Fluids Eng.* **2013**, *135*, 084501. [CrossRef]
5. Litwin, W. Experimental research on water lubricated three layer sliding bearing with lubrication grooves in the upper part of the bush and its comparison with a rubber bearing. *Tribol. Int. A* **2015**, *82*, 153–161. [CrossRef]
6. Cabrera, D.; Woolley, N.; Allanson, D.; Tridimas, Y. Film pressure distribution in water-lubricated rubber journal bearings. *Proc. Inst. Mech. Eng. J J. Eng. Tribol.* **2005**, *219*, 125–132. [CrossRef]
7. Heberley, B.D. Advances in Hybrid Water-Lubricated Journal Bearings for Use in Ocean Vessels. Ph.D. Thesis, Massachusetts Institute of Technology, Cambridge, MA, USA, 2013.
8. Majumdar, B.; Pai, R.; Hargreaves, D. Analysis of water-lubricated journal bearings with multiple axial grooves. *Proc. Inst. Mech. Eng. J J. Eng. Tribol.* **2004**, *218*, 135–146. [CrossRef]
9. Wang, X.; Adachi, K.; Otsuka, K.; Kato, K. Optimization of the surface texture for silicon carbide sliding in water. *Appl. Surf. Sci.* **2006**, *253*, 1282–1286. [CrossRef]
10. Litwin, W. Influence of surface roughness topography on properties of water-lubricated polymer bearings: Experimental research. *Tribol. Trans.* **2011**, *54*, 351–361. [CrossRef]
11. Wang, J.; Chen, B.; Liu, N.; Han, G.; Yan, F. Combined effects of fiber/matrix interface and water absorption on the tribological behaviors of water-lubricated polytetrafluoroethylene-based composites reinforced with carbon and basalt fibers. *Compos. A Appl. Sci. Manuf.* **2014**, *59*, 85–92. [CrossRef]
12. Zhao, B.; Zhang, S.; Man, J.; Zhang, Q.; Chen, Y. A modified normal contact stiffness model considering effect of surface topography. *Proc. Inst. Mech. Eng. J J. Eng. Tribol.* **2015**, *229*, 677–688. [CrossRef]
13. Pai, R.; Pai, R. Stability of four-axial and six-axial grooved water-lubricated journal bearings under dynamic load. *Proc. Inst. Mech. Eng. J J. Eng. Tribol.* **2008**, *222*, 683–691. [CrossRef]
14. Zhu, D.; Wang, Q.J. On the λ ratio range of mixed lubrication. *Proc. Inst. Mech. Eng. J J. Eng. Tribol.* **2012**, *226*, 1010–1022. [CrossRef]
15. Yeo, C.-D.; Katta, R.R.; Lee, J.; Polycarpou, A.A. Effect of asperity interactions on rough surface elastic contact behavior: Hard film on soft substrate. *Tribol. Int.* **2010**, *43*, 1438–1448. [CrossRef]
16. Sahlin, F.; Larsson, R.; Marklund, P.; Almqvist, A.; Lugt, P. A mixed lubrication model incorporating measured surface topography. Part 2: Roughness treatment, model validation, and simulation. *Proc. Inst. Mech. Eng. J J. Eng. Tribol.* **2010**, *224*, 353–365. [CrossRef]
17. Sahlin, F.; Larsson, R.; Almqvist, A.; Lugt, P.; Marklund, P. A mixed lubrication model incorporating measured surface topography. Part 1: Theory of flow factors. *Proc. Inst. Mech. Eng. J J. Eng. Tribol.* **2010**, *224*, 335–351. [CrossRef]
18. Lin, J.-R.; Hsu, C.-H.; Lai, C. Surface roughness effects on the oscillating squeeze-film behavior of long partial journal bearings. *Comput. Struct.* **2002**, *80*, 297–303. [CrossRef]
19. Hsu, T.-C.; Chen, J.-H.; Chiang, H.-L.; Chou, T.-L. Lubrication performance of short journal bearings considering the effects of surface roughness and magnetic field. *Tribol. Int.* **2013**, *61*, 169–175. [CrossRef]
20. He, T.; Zou, D.; Lu, X.; Guo, Y.; Wang, Z.; Li, W. Mixed-lubrication analysis of marine stern tube bearing considering bending deformation of stern shaft and cavitation. *Tribol. Int.* **2014**, *73*, 108–116. [CrossRef]
21. Tala-Ighil, N.; Fillon, M. A numerical investigation of both thermal and texturing surface effects on the journal bearings static characteristics. *Tribol. Int.* **2015**, *90*, 228–239. [CrossRef]
22. Brito, F.P.; Miranda, A.S.; Fillon, M. Analysis of the effect of grooves in single and twin axial groove journal bearings under varying load direction. *Tribol. Int.* **2016**, *103*, 609–619. [CrossRef]
23. Tala-Ighil, N.; Fillon, M.; Chaouche, A.B.; Mokhtari, A. Numerical study of thermal effects in the hydrodynamic behavior of textured journal bearings. *AIP Conf. Proc.* **2015**, *1648*, 850076. [CrossRef]

24. Zhang, H.; Hua, M.; Dong, G.-N.; Zhang, D.-Y.; Chin, K.-S. A mixed lubrication model for studying tribological behaviors of surface texturing. *Tribol. Int.* **2016**, *93*, 583–592. [CrossRef]
25. Vlădescu, S.-C.; Medina, S.; Olver, A.V.; Pegg, I.G.; Reddyhoff, T. Lubricant film thickness and friction force measurements in a laser surface textured reciprocating line contact simulating the piston ring–liner pairing. *Tribol. Int.* **2016**, *98*, 317–329. [CrossRef]
26. Litwin, W.; Dymarski, C. Experimental research on water-lubricated marine stern tube bearings in conditions of improper lubrication and cooling causing rapid bush wear. *Tribol. Int.* **2016**, *95*, 449–455. [CrossRef]
27. Zhang, X.; Yin, Z.; Gao, G.; Li, Z. Determination of stiffness coefficients of hydrodynamic water-lubricated plain journal bearings. *Tribol. Int.* **2015**, *85*, 37–47. [CrossRef]
28. Illner, T.; Bartel, D.; Deters, L. Determination of the transition speed in journal bearings under consideration of bearing deformation. *Tribol. Int. A* **2015**, *82*, 58–67. [CrossRef]
29. Gonçalves, D.; Graça, B.; Campos, A.V.; Seabra, J.; Leckner, J.; Westbroek, R. On the film thickness behaviour of polymer greases at low and high speeds. *Tribol. Int.* **2015**, *90*, 435–444. [CrossRef]
30. Dadouche, A.; Conlon, M.J. Operational performance of textured journal bearings lubricated with a contaminated fluid. *Tribol. Int.* **2016**, *93*, 377–389. [CrossRef]
31. Lu, X.; Khonsari, M.M. An Experimental Investigation of Dimple Effect on the Stribeck Curve of Journal Bearings. *Tribol. Lett.* **2007**, *27*, 169. [CrossRef]
32. Tala-Ighil, N.; Fillon, M.; Maspeyrot, P. Effect of textured area on the performances of a hydrodynamic journal bearing. *Tribol. Int.* **2011**, *44*, 211–219. [CrossRef]
33. Cristea, A.-F.; Bouyer, J.; Fillon, M.; Pascovici, M.D. Transient Pressure and Temperature Field Measurements in a Lightly Loaded Circumferential Groove Journal Bearing from Startup to Steady-State Thermal Stabilization. *Tribol. Trans.* **2017**, *60*, 988–1010. [CrossRef]
34. Xie, Z.; Rao, Z.-S.; Liu, L.; Chen, R. Theoretical and experimental research on the friction coefficient of water lubricated bearing with consideration of wall slip effects. *Mech. Ind.* **2016**, *17*, 106–119. [CrossRef]
35. Xie, Z.L.; Rao, Z.S.; Na, T.; Liu, L. Investigations on transitions of lubrication states for water lubricated bearing. Part I: Determination of friction coefficients and film thickness ratios. *Ind. Lubr. Tribol.* **2016**, *68*, 404–415. [CrossRef]
36. Xie, Z.L.; Rao, Z.S.; Na, T.; Liu, L. Investigations on transitions of lubrication states for water lubricated bearing. Part II: Further insight into the film thickness ratio lambda. *Ind. Lubr. Tribol.* **2016**, *68*, 416–429. [CrossRef]
37. Bungartz, H.-J.; Mehl, M.; Schäfer, M. *Fluid Structure Interaction II: Modelling, Simulation, Optimization*; Springer: Berlin, Germany, 2010.
38. Chung, T.J. *Computational Fluid Dynamics*; Cambridge University Press: Cambridge, UK, 2010.
39. Hamrock, B.J. *Fundamentals of Fluid Film Lubrication*; Marcel Dekker: New York, NY, USA, 2004.
40. Szeri, A.Z. *Fluid Film Lubrication*; Cambridge University Press: Cambridge, UK, 2011.
41. Leighton, M.; Rahmani, R.; Rahnejat, H. Surface specific flow factors for prediction of cross-hatched surfaces. *Surf. Topogr. Metrol. Prop.* **2016**, *4*, 025002. [CrossRef]

coatings

MDPI

Article

Effects of Surface Microstructures on Superhydrophobic Properties and Oil-Water Separation Efficiency

Yangyang Chen [1,2], Shengke Yang [1,2,*], Qian Zhang [1,2], Dan Zhang [1,2], Chunyan Yang [1,2], Zongzhou Wang [1,2], Runze Wang [1,2], Rong Song [1,2], Wenke Wang [1,2] and Yaqian Zhao [3]

[1] Key Laboratory of Subsurface Hydrology and Ecological Effects in Arid Region, Ministry of Education, Chang'an University, Xi'an 710054, China; 2015129085@chd.edu.cn (Y.C.); 2015129090@chd.edu.cn (Q.Z.); 2017129063@chd.edu.cn (D.Z.); 2017129066@chd.edu.cn (C.Y.); 2015229061@chd.edu.cn (Z.W.); 2016129077@chd.edu.cn (R.W.); 2016229036@chd.edu.cn (R.S.); 2017129064@chd.edu.cn (W.W.)
[2] School of Environmental Science and Engineering, Chang'an University, Xi'an 710054, Shaanxi, China
[3] Dooge Centre for Water Resource Research, School of Civil Engineering, University College Dublin, Belfield, Dublin 4, Ireland; 2016229046@chd.edu.cn
* Correspondence: yskfxh@chd.edu.cn or ysk110@126.com; Tel.: +86-29-8558-5589

Received: 16 December 2018; Accepted: 22 January 2019; Published: 24 January 2019

Abstract: In order to explore the effects of microstructures of membranes on superhydrophobic properties, it is critical, though, challenging, to study microstructures with different morphologies. In this work, a combination of chemical etching and oxidation was used and some copper meshes were selected for grinding. Two superhydrophobic morphologies could be successfully prepared for oil-water separation: a parabolic morphology and a truncated cone morphology. The surface morphology, chemical composition, and wettability were characterized. The results indicated that the water contact angle and the advancing and receding contact angles of the parabolic morphology were 153.6°, 154.6° ± 1.1°, and 151.5° ± 1.8°, respectively. The water contact angle and the advancing and receding contact angles of the truncated cone morphology were 121.8°, 122.7° ± 1.6°, and 119.6° ± 2.7°, respectively. The separation efficiency of the parabolic morphology for different oil-water mixtures was 97.5%, 97.2%, and 91%. The separation efficiency of the truncated cone morphology was 93.2%, 92%, and 89%. In addition, the values of the deepest heights of pressure resistance of the parabolic and truncated cone morphologies were 21.4 cm of water and 19.6 cm of water, respectively. This shows that the parabolic morphology had good separation efficiency, pressure resistance, and superhydrophobic ability compared with the truncated cone morphology. It illustrates that microstructure is one of the main factors affecting superhydrophobic properties.

Keywords: superhydrophobic materials; rough morphology; parabolic morphology; truncated cone morphology; oil-water separation

1. Introduction

Both oil leakage accidents and arbitrary discharges of oily wastewater have caused great damage to water resources [1–4]. In order to recycle crude oil that has leaked into the water and to purify the water [5], a lot of efforts have been made to separate the oils from the water surface, including using oil skimmers, flotation, and gravity separation. However, most of them have some drawbacks, such as low separation efficiency, low flux, and high operation cost [6,7]. In recent years, inspired by surface structures like the lotus leaf self-cleaning surface and mosquito compound eyes [8], a series of superhydrophobic materials have been prepared and used for oil-water separation [9–12]; moreover, great advances have been achieved and this has attracted extensive attention from scholars.

Previous studies have shown that superhydrophobic properties are determined by surface energy; simultaneously, the water contact angle, and the advancing and receding contact angles are important indexes to revealing superhydrophobic properties [13–18]. In order to represent superhydrophobic properties at a deeper level, S. Hoshian [19] and G. McHale [18] studied advancing and receding contact angles to better illustrate their importance and the superhydrophobic properties of membranes. Meanwhile, scholars continued to construct theoretical models in order to explore the theoretical relationship between all contact angles and superhydrophobic properties. The Wenzel model [20] was classical and it considered the rough surface to be grooved. However, it was different from the phenomenon of the actual superhydrophobic surface with lotus effect. The Wenzel model cannot explain the existence of droplets on a superhydrophobic solid surface. Yamamoto et al. [21] only thought that microscopic rough surfaces were pillar surfaces and did not consider the case where the protrusive surfaces were curved. Nosonovsky et al. [22] studied the relationship between surface roughness and the wetting properties of the sawtooth structures, periodic rough structures, rectangular structures, hemispherically topped cylindrical structures, conical structures, and pyramidal structures, and found that both hemispherically topped cylindrical structures and pyramidal structures can achieve a stable Cassie wetting state, which is consistent with the fact that the top surface of microscopic rough surfaces is often spherical or parabolic in the actual application. However, previous research work has lacked a more in-depth discussion of this situation. Eyal Bittoun and Abraham Marmur [23] constructed four different morphologies of rough surfaces: a cylinder, a truncated cone, a paraboloid and a hemisphere. They made a theoretical model analysis of wettability area for four different protrusive structural models in order to study the relationship between four different protrusive structural models and surface hydrophobicity. The conclusion was that parabolic protrusions seemed to be most advantageous, with the truncated cone second. However, the influence of morphology on surface hydrophobicity was only a theoretical prediction. It was necessary to verify the theoretical prediction results by fabricating superhydrophobic materials with different morphologies such as a truncated cone morphology and a parabolic rough morphology.

In addition, the correct choice of substrate materials is a key factor in achieving superhydrophobicity. Therefore, researchers have used different substrate materials: graphene sponge [24], poly vinylidene fluoride (PVDF) aerogel [25], zeolite [26], silicon [27], and copper [28], etc. to fabricate superhydrophobic surfaces. Superhydrophobic sponges can be prepared by the simple treatment of commercially available Pu-sponges with superhydrophobic nanoparticles [29]. Superhydrophobic surfaces can be created via surface modification of polymers [30]. However, the synthesis of graphene sponges involves multistep procedures, restricting their large-scale production for practical applications [31,32]. With PVDF aerogel it is difficult to remove the absorbed oil quickly, reducing the recyclability [33,34]. For absorbent materials like zeolite, it was difficult to recover oil, and material durability was poor, etc. [35]. The major drawback of silicon-based surfaces is mechanical fragility [19]. Copper meshes have attracted widespread attention for their mechanical strength, low density, high specific surface, and environmentally friendliness [36,37]. The as-prepared copper materials are promising for the preparation of superhydrophobic materials [38].

In addition, to fabricate superhydrophobic property surfaces, a rough structure must be modified with low surface energy coatings [39,40]. Methods reported include: chemical vapor deposition [41], hydrothermal [42], electro spinning [43], and etching [44]. However, many scholars have further improved the superhydrophobicity of the surfaces of copper meshes by etching and oxidation [38] but have not paid attention to variation in surface morphologies. It is significant to obtain superhydrophobic surfaces with different morphologies by etching and oxidation.

In recent years, further research has indicated that different pretreatment of the substrate material can obtain superhydrophobic materials with different morphologies and superior performance. Pan et al. [45] pretreated copper meshes with acetone, ultrapure water, HCl, nitric acid, methanol, etc. Maryam Khosravi et al. [46] pretreated stainless steel meshes with acetone, deionized water, $FeCl_3$, ethanol, carbon soot (CS), pyrrol liquid, etc. Rong et al. [25] pretreated copper foam meshes with diluted HCl, ethanol, deionized water, ammonium persulfate ($(NH_4)_2S_2O_8$), dibasic sodium phosphate (Na_2HPO_4), etc. Elmira Velayi et al. [6] pretreated ZnO with deionized water, hexamethylenetetramine (HMTA), ethanol, distilled water, etc. Although different materials and pretreatment methods had been used to prepare rough surfaces, the rough structures had not been described in detail and associated with theoretical models.

In this paper, copper meshes were used as substrate materials. The copper meshes were pretreated through a series of processes including etching, oxidation, grinding, and modification. A parabolic morphology and a truncated cone morphology were obtained. Finally, the two kinds of meshes' membranes were characterized respectively and applied to an oil-water separation. The hydrophobic properties of the superhydrophobic materials with different morphologies and their oil-water separation efficiency were verified.

2. Experimental

2.1. Materials

Copper meshes (200 meshes) were purchased from Shenyang Copper Network Co., Ltd. (Shenyang, China). Acetone (purity > 99%), ethanol (purity = 99.5%), benzene (purity > 99%), stearic acid (SA), carbon tetrachloride (purity > 99%), $FeCl_3$ (35 wt %) (purity > 98%) and H_2O_2 (30 wt %) were purchased from Tianjin Kemiou Chemical Reagent Co., Ltd (Tianjin, China). Engine oil was purchased from Hubei Jianyuan Chemical Co., Ltd (Wuhan, China). SiC papers (800 meshes) were purchased from Shanghai Chaowei Nano Technology Co., Ltd (Shanghai, China). The experimental water was deionized water. All other chemicals were analytical-grade reagents.

2.2. Fabrication of Superhydrophobic Materials

Copper meshes (200 meshes) that were cut into sizes of 4 cm × 4 cm were tilted into a beaker. The copper meshes were then cleaned ultrasonically in acetone, ethanol, and deionized water for 10 min to remove oil and inorganic impurities on their surfaces. The cleaned copper meshes that were placed flat on the configured 35% $FeCl_3$ solution were washed and etched in an ultrasonic bath for 20 min. The etched copper meshes were placed in 30% H_2O_2 solution and etched ultrasonically for 1 min. The as-prepared copper meshes surfaces were ground 50 times with SiC papers. The as-prepared copper meshes were then immersed in a 10 mol/L ethanolic solution of SA (with a volume ratio of 1:3, which refers to the volume ratio of ethanol to water) at 60 °C for 30 h. All experiments were performed at room temperature. Finally, the copper samples were successively rinsed with ethanol and deionized water and dried in air at room temperature and sealed for use. In order to better illustrate the preparation process of the two morphologies, the preparation process has been represented by a schematic diagram, as shown in Figure 1.

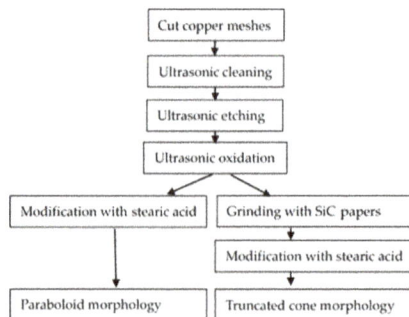

Figure 1. Preparation process schematic diagram of the two morphologies.

2.3. Oil-Water Separation Test

The as-prepared copper meshes that were trimmed into dimensions of 4 cm × 4 cm were fixed on a self-made separation apparatus. The oil/water separation performance of the copper meshes was determined by gravity-driven and capillary force-driven oil/water mixture separation experiments. An oil/water mixture of benzene-water, carbon tetrachloride-water, and engine oil-water was used for the separation experiments. For convenience, deionized water was stained with methylene blue. Oils were stained with Sudan I.

2.4. Characterization

2.4.1. Surface Morphology Characterization

The surface morphologies of the SA-coated copper meshes obtained in different etching solutions were characterized using a microscope (LW300LFT, Beijing Taike Instrument Co., Ltd. Beijing, China) together with recording SEM imaging on a FEI Quanta 200 SEM (FEI company, Hillsboro, OR, USA) to confirm the successful formation of the microstructure. The chemical composition of the copper mesh surface modified with SA was analyzed via Fourier transform infrared spectroscopy (FT-IR) (Magna-IR 560, Nicolet, Beijing, China) and X-ray photoelectron spectroscopy (XPS) (AXIS ULTRA, Kratos Analytical Ltd., Kyoto, Japan)

2.4.2. Contact Angle Measurements

The water contact angles (WCAs), oil contact angles, and advancing and receding contact angles of the samples were measured at ambient temperature using a SL200KS contact angle meter (American Kono Industrial Co. Ltd., Seattle, WA, USA); the volume of the test water droplets and oil droplets were approximately 3 μL. At least three parallel positions on the surface were measured to obtain average contact angle values. The advancing and receding contact angles were determined by tilting experiments.

3. Results and Discussion

3.1. Surface Microstructure after Ultrasonic Etching and Oxidation

Surfaces of the copper meshes with different roughnesses were obtained by etching with 35% $FeCl_3$ and oxidation with 30% H_2O_2 solution. In order to further observe the surface microstructure of the copper meshes, however, since the microscope could only roughly see the differences in surface morphology of the copper meshes, the copper meshes were characterized by SEM. Changes in the surface of the copper meshes were observed by microscope and SEM, respectively.

Figure 2 shows microscope and SEM images of the copper meshes before and after etching and oxidation. As shown in Figure 2a, the size of the copper meshes that were untreated was uniform and the thickness of the copper meshes was basically the same. Figure 2b shows that the diameter of part of the etched and oxidized copper mesh wires was slightly narrowed. Figure 2c,d are SEM images under 800× magnification. Comparing Figure 2c,d, it can be seen that the surface of the untreated copper meshes was smooth. The etched and oxidized copper meshes had no change substantially in shape, but the surface became rough. Moreover, the average sizes of the meshes shown in Figure 2c,d were determined as 71 and 70 μm, respectively, by measuring 10 times randomly. The diameter of the copper mesh wire was slightly narrowed. It was basically consistent with what was observed under the microscope, indicating that after ultrasonic etching using the acidic etchant $FeCl_3$ solution and ultrasonic oxidation using the H_2O_2 solution, the copper meshes were subjected to cavitation impact and the surfaces of copper substrates were deformed and etched, which confirmed that cavitation had occurred [45]. Moreover, the surfaces of the copper substrates became rough, which proved that the etching and oxidation processes increased the surface roughness to a certain extent [38,47,48].

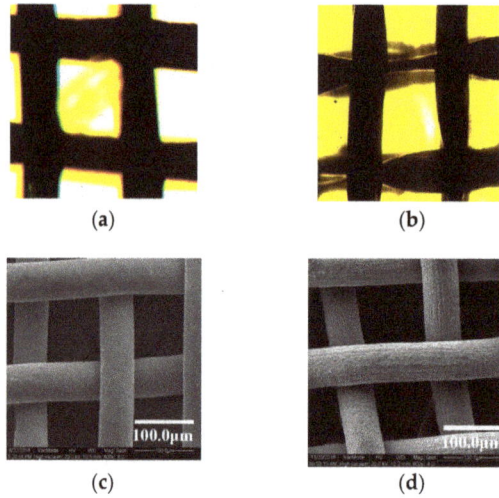

Figure 2. Copper mesh morphologies under microscope and SEM: (**a**,**c**) untreated; (**b**,**d**) etched and oxidized.

3.2. Surface Microstructure by Grinding with SiC

In order to change the surface morphologies of the copper meshes, several steps were performed by etching with 35% $FeCl_3$, using oxidation with 30% H_2O_2 solution, and grinding with SiC. Figure 3 shows microscope and SEM images of those etched and oxidized copper meshes that were ground with SiC and without SiC.

Comparing Figure 3a,b, there were some differences in the surface morphologies of the copper meshes before and after grinding with SiC. As in Figure 3a, the unground copper mesh surfaces were even. As in Figure 3b, the edges of the copper meshes that were ground with SiC were uneven and the surface became rough. It can be seen via microscope that the surface of copper meshes ground with SiC can become rough [49]. To deeply understand the surface morphologies of as-obtained samples before and after grinding, SEM analysis was conducted on copper meshes. Figure 3c,d are SEM images under 5000× magnification. In Figure 3c, the surface of the mastoid structure of the copper meshes was smooth and the papillae were regularly parabolic or granular protrusions. In Figure 3d, the papillae structures of the copper meshes had distinct angular structures and were arranged in a uniform and orderly manner. It was observed by SEM that the surface of parabolic or granular protrusions had been flattened after grinding by SiC, indicating that certain morphology features had appeared on the surface of the copper meshes before and after grinding.

Figure 3. *Cont.*

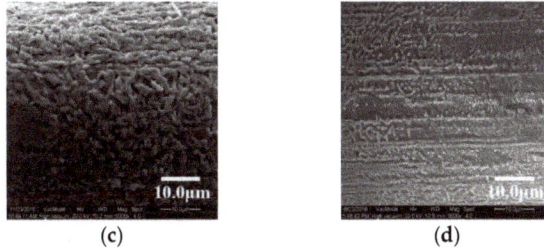

Figure 3. Copper mesh morphology after etching and oxidation under microscope and SEM: (a,c) unground; (b,d) ground.

3.3. Surface Microstructure by Modification with SA

Previous studies have shown that using SA as modifier can form a superhydrophobic membrane surface on the surface of the materials [50–52]. Different microstructures can be obtained by etching with 35% FeCl$_3$, oxidation with 30% H$_2$O$_2$ solution, grinding with SiC, and modification with SA. Thus, the surface microstructures of the copper meshes that were modified with SA were observed via microscope and SEM.

Figure 4 shows the microscope and SEM images of the treated copper meshes that were modified with SA. In Figure 4a, it can be clearly seen that the surface morphologies of the copper meshes were still those of porous structures and had not been completely covered by SA, and only the pore size of the copper meshes had been reduced. In Figure 4b, a layer of SA on copper meshes was clearly evident.

SEM images were performed to further prove the successful formation of SA-decorated copper meshes. In Figure 4c, as can been seen from the fact that the surface structures of the copper meshes were comprised of many microprotrusions, and that many petallike clusters were present on the surface of the copper mesh surfaces, there are obvious spacings and gaps between the mastoid clusters which can capture a lot of air, demonstrating that air pockets were necessary to achieve surface superhydrophobicity [25,53,54]. The morphologies of many petallike papilla clusters were parabolic, arranged uniformly and with height kept consistent. Furthermore, air trapped under the space of protrusion clusters can cause the presence of air cushions on the surface of membranes which can prevent contact between water droplets and the solid surface, thereby increasing the contact angle of the water droplets [38,54]. Figure 4c shows that the copper meshes can form a rough structure similar to a paraboloid. In Figure 4d, a number of blocky and flaky bulges are seen to have arisen on the surface and the blocky mastoids formed by agglomeration of these protrusive structures appear as columnar, as if a series of rough structures with a uniform height and a truncated cone had appeared on the surface of the copper substrates. Figure 4d shows that the copper meshes can form a rough structure similar to a truncated cone.

In order to prove that the two morphologies obtained from Figure 4 are a paraboloid and a truncated cone, respectively, size fitting curves that are drawn by SEM images 4c,d are shown in Figure 4e,f. It can be clearly seen from Figure 4e,f that the surface morphologies of the unground copper meshes were similar to a paraboloid and the surface morphologies of the copper meshes that were ground with SiC were similar to a truncated cone.

Figure 5 shows the FT-IR pattern of the stearic acid and prepared hydrophobic surfaces, respectively. Figure 5a indicates the FT-IR pattern of stearic acid. The spectrum of stearic acid exhibits adsorption peaks at about 2917 and 2852 cm^{-1} in the high frequency region, which, respectively, are attributed to C–H asymmetric and symmetric stretching vibrations. Moreover, due to the presence of carboxyl groups in stearic acid, a characteristic peak appears at 1707 cm^{-1}. The peak at 1707 cm^{-1} represents the stretch absorption of the C=O bond. Figure 5b is the FT-IR pattern of a sample soaked with 10 mmol/L of stearic acid in an ethanol/water solution for 30 h. When the copper mesh was immersed in the stearic acid ethanol/water solution, the same two peaks as for stearic acid appeared at 2917 and 2852 cm^{-1}. The surface of the modified copper mesh showed new characteristic peaks at 1720 and 1590 cm^{-1}. The two adsorption peaks may stem from asymmetric and symmetric stretches of carboxyl groups. Based on the analysis of the FT-IR spectrum, it can be seen that the surface of the copper meshes has been successfully modified with SA.

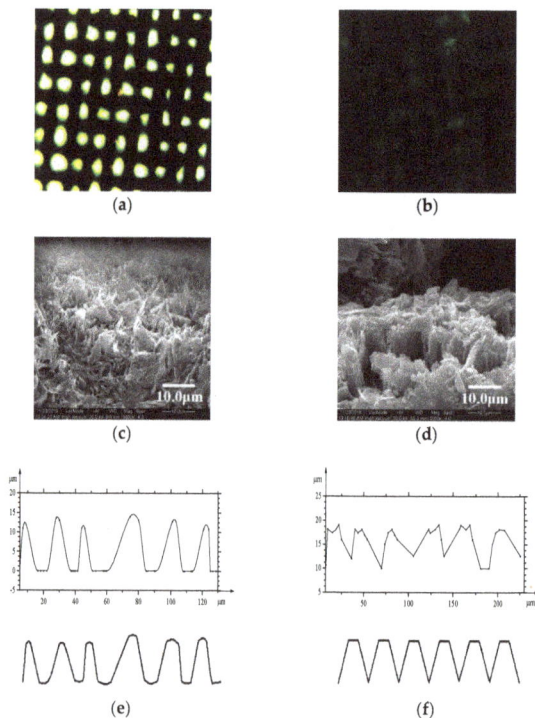

Figure 4. Copper mesh morphology after etching, oxidation, and modification with SA under microscope and SEM: (**a,c**) unground; (**b,d**) ground; (**e**) the fitting curve of Figure 4c; (**f**) the fitting curve of Figure 4c.

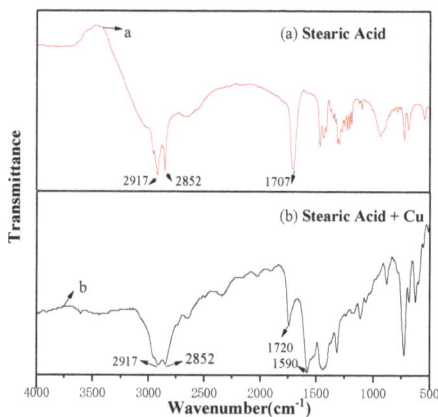

Figure 5. Fourier transform infrared spectroscopy (FT-IR) spectra: (**a**) stearic acid; (**b**) stearic acid + Cu.

To investigate the properties of superhydrophobic membranes, the surface composition was characterized by X-ray electron spectroscopy. Figure 6a shows the XPS survey spectrum of the as-prepared superhydrophobic copper surface. Figure 6b,c are high-resolution spectra of Cu 2*p* and C 1*s* respectively. As can be seen from Figure 6a, the peaks of oxygen, carbon and Cu appeared on the survey spectrum. The presence of C 1*s* and O 1*s* indicated the presence of SA in the membranes. However, the peak of Cu was not very obvious, probably because the coating of SA was thicker (100–300 μm), resulting in a weaker peak signal of Cu. It can be seen from Figure 6b that the main Cu $2p_{3/2}$ and $2p_{1/2}$ can be observed at the binding energies around 933.56 and 953.02 eV, respectively. As can be seen from Figure 6c, the C–C bond, the C–O bond, and the O=C–O bond appeared at 284.8, 285.63, and 289.11 eV, respectively. Moreover, a weak Cu–C bond was formed at 290.3 eV. Due to the presence of this weak bond, the copper meshes and SA were more closely linked together. It can be seen from Figure 6b,c that there was Cu^{2+} on the Cu 2*p* peak and there was COO– on the C 1*s* peak; thus, these results indicate that SA was well coated on the copper mesh membrane.

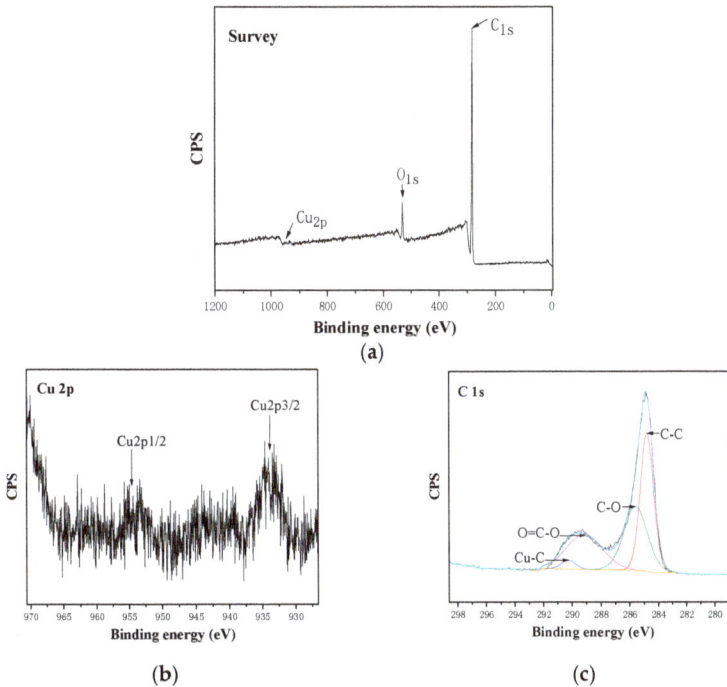

Figure 6. X-ray electron spectroscopy (XPS) spectra of superhydrophobic copper modified in an ethanol solution of SA for 30 h: (**a**) survey; (**b**) Cu 2*p* spectra; (**c**) C 1*s* spectra.

3.4. Wettability of the Copper Meshes

The water contact angle, advancing contact angle (θ_A) and receding contact angle (θ_R) are critical to the characterization of superhydrophobic surfaces. Figure 7 shows the WCA and the advancing and receding contact angles of the parabolic and truncated cone morphologies. In Figure 7a–c, the WCA and the advancing and receding contact angles of the parabolic morphology were 153.6°, 154.6° ± 1.1°, and 151.5° ± 1.8°, respectively, which showed that the parabolic morphology became a superhydrophobic surface after treatment with SA. In Figure 7b–d, the WCA and advancing and receding contact angles of the truncated cone morphology were 121.8°, 122.7° ± 1.6°, and 119.6°

± 2.7°, respectively, which showed that the truncated cone morphology became a hydrophobic surface after treatment with SA. However, it was not a superhydrophobic surface.

Cao et al. [38] have synthesized copper meshes that were modified by n-dodecanethiol (DDT), and the surface of the copper meshes measured a static WCA of 152°, which showed that it was very important to grasp the surface structure and manufacturing process for superhydrophobic properties. Yuan et al. [55] synthesized copper meshes via a combination of polydimethylsiloxane (PDMS) template and etching method which produced a rough surface similar to a lotus leaf shape, upon which the surface of the copper meshes modified with SA measured a static WCA of 131°, demonstrating that it had a smaller contact angle than for the copper meshes etched and oxidized ultrasonically, which confirms that with ultrasonic etching and oxidation, the as-prepared meshes with nano-wall surface structures show good superhydrophobicity.

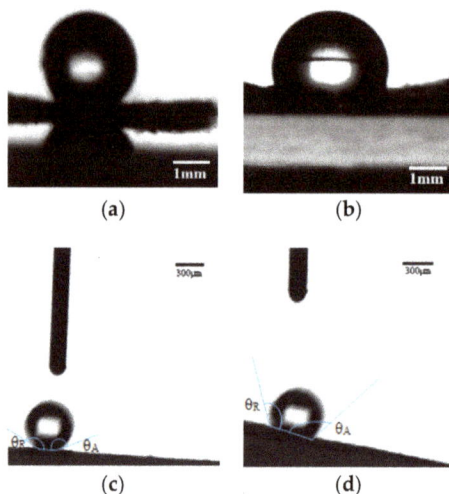

Figure 7. Wettability of the surfaces of the mesh membranes; water droplets on: (**a,c**) parabolic morphology; (**b,d**) truncated cone morphology.

The wetting properties of the as-prepared copper meshes towards oil (for engine oil and carbon tetrachloride, the surface tension of engine oil is 0.04 N/m and the surface tension of carbon tetrachloride is 0.02683 N/m [56])were evaluated through static contact angle. Figure 8a,b show the wettability images of engine oil on the paraboloid and truncated cone morphologies, respectively. Figure 8c,d show the wettability image of carbon tetrachloride on the paraboloid and truncated cone morphologies, respectively.

In Figure 8a, after three measurements being performed and averaged, the surface of the copper meshes produced a CA of 5°, which showed that the structure of the paraboloid became a superoleophilic surface. In Figure 8b, the surface of the copper meshes produced a CA of 32.9°, which showed that the structure of the truncated cone became a lipophilic surface. However, it was not a superoleophilic surface. Figure 8c,d shows that the viscosity of carbon tetrachloride was lower than engine oil and thus, the carbon tetrachloride could easily spread onto the copper mesh surfaces as carbon tetrachloride dropped on the copper meshes. The surface of the copper meshes produced a CA of 0.1°, which indicates that the microstructures of the paraboloid and truncated cone were superoleophilic. By comparing Figure 8a,c, and Figure 8b,d, it can be recognized that when the oil types were different (for example, engine oil/carbon tetrachloride), the contact angles on the surface of the material were different.

The wettability of water droplets, engine oil, and carbon tetrachloride on the paraboloid and truncated cone was measured, confirming that the microstructures of the surfaces had great influence on the CA and hydrophobic and lipophilic properties of materials [57,58]. Moreover, the different types of oil also led to differences in the CA on the material's surface. In summary, the rough structure of the paraboloid was superhydrophobic and superoleophilic, demonstrating it was more beneficial to achieve superhydrophobicity and superoleophilicity. The rough structure of the truncated cone was hydrophobic and lipophilic; however, it was not a superhydrophobic surface, which confirms that special surface morphologies are important factors that have great influence on surface superhydrophobicity [59,60].

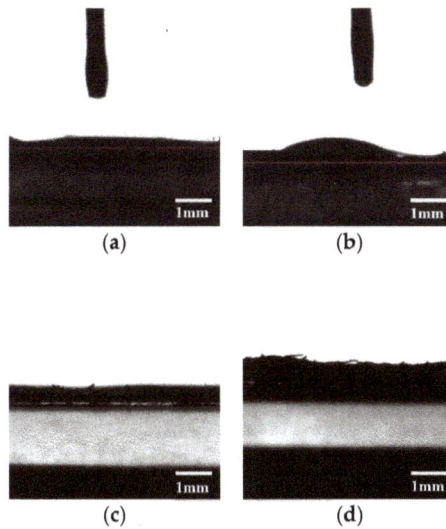

Figure 8. Wettability of the surfaces of the mesh membranes; oil droplets on: (**a,c**) parabolic morphology; (**b,d**) truncated cone morphology.

To better illustrate the contact properties of water droplets and various oil droplets on the two morphologies, all the contact angles are listed in Table 1.

Table 1. Contact angles of different materials on different morphologies.

Different Situations	Different Types of Contact Angles	Angel	
		Parabolic	Truncated Cone
water	WCA	153.6°	121.8°
	θ_A	154.6° ± 1.1°	151.5° ± 1.8°
	θ_R	122.7° ± 1.6°	119.6° ± 2.7°
Engine oil	–	5°	32.9°
Carbon tetrachloride	–	0.1°	0.1°

3.5. Oil-Water Separation Performance

In order to understand the oil-water separation performance of the two morphologies, parabolic morphology was taken as an example. Carbon tetrachloride-water and benzene-water were selected for oil-water separation experiments. Deionized water was dyed with methylene blue, and carbon tetrachloride and benzene were dyed with Sudan I. The volumes of deionized water, engine oil and carbon tetrachloride used for the experiments were both 30 mL. The volume of benzene was 45 mL.

The separation experimental processes of benzene-water are shown in Figure 9a. The density of benzene was less than water and benzene passed first through the top glass tube and then flowed into the segregation apparatus. Due to the hydrophobic and lipophilic properties of the membranes, benzene passed through the surface of the membranes and flowed into the beaker. The deionized water was repelled on the membranes. The entire separation process took about 2 min. Engine oil underwent the same process as benzene.

The separation experimental processes of carbon tetrachloride-water are shown in Figure 9b. The density of carbon tetrachloride was greater than water and the deionized water passed first through the top glass tube and then flowed into the segregation apparatus. Due to the hydrophobic and lipophilic properties of the membranes, the deionized water was repelled by the membranes while carbon tetrachloride passed through the deionized water, reached the surface of the membranes, and flowed into the beaker quickly. The entire separation process took about 2 min.

When the separation process was completed, the oils in the beaker were poured into a measuring cylinder and the volumes of oils were measured after separation. It was found that the volumes of benzene, carbon tetrachloride and engine oil after separation using the parabolic morphology were approximately 43.8, 29.2, and 27.3 mL, respectively. The volumes of benzene, carbon tetrachloride, and engine oil after separation via the truncated cone morphology were approximately 41.7, 27.6, and 26.7 mL, respectively.

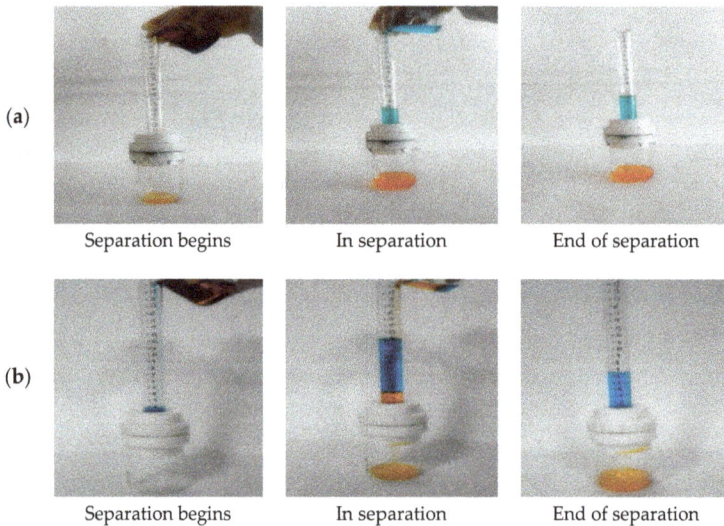

Figure 9. Separation effect diagram of oil-water mixture using parabolic membranes: (**a**) benzene-water; (**b**) carbon tetrachloride-water.

According to the above method a mixture of benzene, carbon tetrachloride, engine oil, and deionized water was supplied to oil-water separation experiments on the surface of the truncated cone. In order to assess the quality of separation of two different microstructures of the copper meshes, after separation of oil and water, according to the amount of oil passing through the filter, the separation efficiency, η (%), was calculated using Equation (1), which is

$$\eta = V_R/V_0 \times 100\% \tag{1}$$

where V_0 is the oil mass of the original oil-water mixture and V_R is the oil mass after separation. The results obtained are shown in Figure 10.

It can be seen that the separation efficiency of the as-prepared rough morphology of the paraboloid was above 91% for three different oil-water mixtures of benzene, carbon tetrachloride, and engine oil. What was more important was that the separation efficiency of benzene and carbon tetrachloride was about 97%. However, given that the separation efficiency of the mixtures of benzene-water, carbon tetrachloride-water and engine oil-water was 93.2%, 92%, and 89% on truncated cone morphologies, it can be found that the oil-water separation efficiency of the rough morphology of the paraboloid was higher than for the truncated cone.

It can be recognized that due to the different types of oil, the contact angles of oil droplets on the surfaces of the copper meshes were different, so the separation efficiency was different [46,54]. The oil-water mixture separation efficiency of the rough morphology of the paraboloid was higher than that of the oil-water mixture with the rough morphology of the truncated cone, demonstrating that superhydrophobicity and oleophilicity facilitate the separation of mixtures of oil and water [54,61]. Moreover, taking the benzene-water mixture as an example, after five tests using the mixture of benzene and water, it was found that the oil-water separation efficiency of the parabolic morphologies was 96% ± 2% and the oil-water separation efficiency of the truncated cone morphologies was 93% ± 1%.

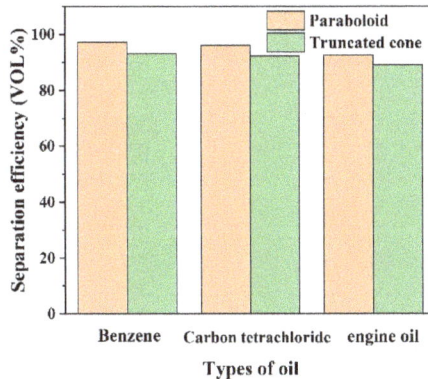

Figure 10. Results of separation efficiency for different oil-water mixtures.

3.6. The Height of Pressure Resistance Study

The height of the pressure resistance of the meshes has a great influence on oil-water separation. For the same membranes, the higher the level of the liquid, the greater the pressure is and the faster the separation rate, though separation quality may decline a little. When the pressure reaches a critical state, the water will pass through. In order to improve the separation rate, it is required that the membranes are used at the deepest height of pressure resistance as much as possible. A device diagram of highest pressure resistance is shown in Figure 11a. The deionized water was slowly added from the upper of the glass tube along the wall of the glass tube. As the height of the liquid column increased, the deionized water was slowly added using a plastic dropper. When the first water droplet flowed into the segregation apparatus, the height of the liquid column at that moment was recorded. The above experiment was repeated several times to find a stable value of the height of the pressure resistance.

The height of the pressure resistance pattern of the paraboloid and truncated cone morphologies are shown in Figure 11b,c. The height of the pressure resistance value of the copper meshes of paraboloid and truncated cone morphologies was 12 cm of water and 6.9 cm of water, respectively.

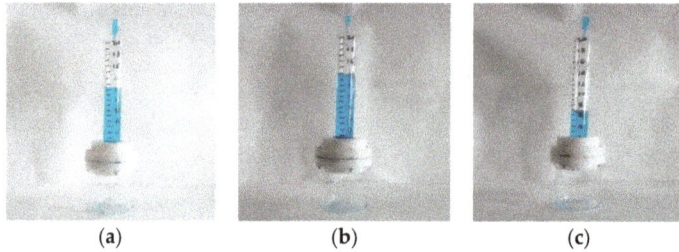

(a)   (b)   (c)

Figure 11. (a) The device used for highest pressure resistance; (b) apparatus of pressure resistance performance test for the parabolic morphology; (c) apparatus of pressure resistance performance test for the truncated cone morphology.

To further describe the separation mechanism of oil and water for the SA-modified meshes, SEM was used to observe the mesh membranes with morphologies of the paraboloid and truncated cone, as shown in Figure 12. Jiang et al. [62] have showed that the deepest height of pressure resistance value of superhydrophobic and superlipophilic membranes by capillary phenomenon, the deepest height of pressure resistance, h, may be calculated using Equation (2), which is

$$h = -2\gamma_{H_2O}\cos\theta/\rho gR \tag{2}$$

where γ_{H_2O} is the interfacial tension of liquid, $\gamma_{H_2O} = 72.75$ mN/m, θ represents the water and membrane CA, ρ_{H_2O} is the density of water (1 g/cm^3), g represents gravitational acceleration (9.8 kg/N), and R is the diagonal length of rectangular meshes.

Copper meshes with rough surfaces which have paraboloid and truncated cone morphologies are shown in Figure 12. In Figure 12a, the copper meshes were covered with a layer of SA. Therefore, the mesh hole was roughly rectangular and the size of the mesh hole was on average about 58 μm × 23 μm; the diagonal length R of the rectangle mesh hole was about 62 μm. The deepest height of pressure resistance, h_1 (cm), was calculated using Equation (2): The deepest height of pressure resistance h_1 of the meshes was 21.4 cm of water. In Figure 12b, the mesh hole was roughly square and the size of the mesh hole was on average about 28 μm × 28 μm; the diagonal length R of the square mesh hole was about 40 μm. The deepest height of pressure resistance, h_2 (cm), was calculated using Equation (2): the deepest height of pressure resistance h_2 of the meshes was 19.6 cm of water. It was smaller than h_1.

It can be seen that both the theoretical value and the actual calculated value prove that the deepest height of pressure resistance of the parabolic copper mesh was significantly larger than that of the truncated cone meshes, which confirmed that the membranes with a parabolic rough morphology had superior superhydrophobicity and deepest height of pressure resistance compared to those with a truncated cone morphology.

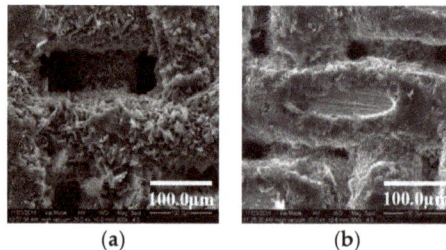

(a)   (b)

Figure 12. SEM of the copper meshes: (a) parabolic morphology; (b) truncated cone morphology.

In order to clearly illustrate the difference between the heights of pressure resistance of the different morphologies, the results are listed in Table 2.

Table 2. The values of the height of pressure resistance under different morphologies.

Different Situations	Height of Pressure Resistance (cm of Water)	
	Parabolic	Truncated Cone
Theoretical value	12	6.9
Actual value	21.4	19.6

4. Conclusions

Two different hydrophobic surfaces on copper meshes were successfully formed by chemical etching, oxidation, and grinding. After modification with SA, surfaces with parabolic morphology exhibited superhydrophobicity. Surfaces with truncated cone morphology showed hydrophobicity.

The WCA and the advancing and receding contact angles of the parabolic morphology were about 153.6°, 154.6° ± 1.1°, and 151.5° ± 1.8°, respectively. The WCA and the advancing and receding contact angles of the truncated cone morphology were about 121.8°, 122.7° ± 1.6°, and 119.6° ± 2.7°, respectively.

The theoretical height of pressure resistance of the parabolic morphology was 12 cm of water and the actual value was 24 cm of water. The theoretical height of pressure resistant of the truncated cone morphology was 6.9 cm of water and the actual value was 19.6 cm of water. This showed that the parabolic morphology had a higher height of pressure resistance than the truncated cone morphology.

The most important thing was to study the separation efficiency of three kinds of water-oil mixtures on the two morphologies. It was found that the parabolic morphology had higher separation efficiency for different oil-water mixtures than the truncated cone morphology. That the parabolic morphology was more superhydrophobic than the truncated cone morphology was confirmed by analyzing the CA, the height of pressure resistance, and separation efficiency. This result illustrates that microstructure is one of the main factors affecting superhydrophobic properties.

Author Contributions: Conceptualization, Y.C.; Data Curation, Z.W. and R.S.; Funding Acquisition, S.Y.; Investigation, Q.Z.; Methodology, Q.Z.; Project Administration, S.Y.; Software, C.Y.; Validation, D.Z.; Visualization, R.W.; Writing—Original Draft, Y.C.; Writing—Review and Editing, S.Y., W.W. and Y.Z.

Funding: This research was funded by the National Key Research and Development Program of China (No. 2016YFC0400701 and 2018YFC0406504), the National Natural Science Foundation of China (No. 41672224) and Henan province transportation science and technology project (No. 2017J4-1).

Acknowledgments: We are highly grateful to the anonymous reviewers for their valuable comments. We also acknowledge the endless support from the staff of Shaanxi Normal University, Xi'an, China.

Conflicts of Interest: The authors declare no conflict of interest.

References

1. Kirby, M.F.; Law, R.J. Accidental spills at sea–risk, impact, mitigation and the need for co-ordinated post-incident monitoring. *Mar. Pollut. Bull.* **2010**, *60*, 797–803. [CrossRef] [PubMed]
2. Li, Z.T.; Lin, B.; Jiang, L.W.; Lin, E.C.; Chen, J.; Zhang, S.J.; Tang, Y.W.; He, F.A.; Li, D.H. Effective preparation of magnetic superhydrophobic Fe$_3$O$_4$/Pu sponge for oil-water separation. *Appl. Surf. Sci.* **2018**, *427*, 56–64. [CrossRef]
3. Kota, A.K.; Kwon, G.; Choi, W.; Mabry, J.M.; Tuteja, A. Hygro-responsive membranes for effective oil-water separation. *Nat. Commun.* **2012**, *3*, 1025. [CrossRef] [PubMed]
4. Crick, C.R.; Bhachu, D.S.; Parkin, I.P. Superhydrophobic silica wool—A facile route to separating oil and hydrophobic solvents from water. *Sci. Technol. Adv. Mater.* **2014**, *15*, 065003. [CrossRef] [PubMed]
5. Crick, C.R.; Ozkan, F.T.; Parkin, I.P. Fabrication of optimized oil–water separation devices through the targeted treatment of silica meshes. *Sci. Technol. Adv. Mater.* **2015**, *16*, 055006. [CrossRef] [PubMed]

6. Velayi, E.; Norouzbeigi, R. Synthesis of hierarchical superhydrophobic zinc oxide nano-structures for oil/water separation. *Ceram. Int.* **2018**, *44*, 14202–14208. [CrossRef]

7. Zhu, H.; Guo, Z. Understanding the separations of oil/water mixtures from immiscible to emulsions on super-wettable surfaces. *J. Bionic Eng.* **2016**, *13*, 1–29.

8. Gao, X.; Yan, X.; Yao, X.; Xu, L.; Zhang, K.; Zhang, J.; Yang, B.; Jiang, L. The dry-style antifogging properties of mosquito compound eyes and artificial analogues prepared by soft lithography. *Adv. Mater.* **2007**, *19*, 2213–2217. [CrossRef]

9. Zhang, Z.; Ge, B.; Men, X.; Li, Y. Mechanically durable, superhydrophobic coatings prepared by dual-layer method for anti-corrosion and self-cleaning. *Colloid Surf. A* **2016**, *490*, 182–188. [CrossRef]

10. Gao, X.; Guo, Z. Mechanical stability, corrosion resistance of superhydrophobic steel and repairable durability of its slippery surface. *J. Colloid Interface Sci.* **2018**, *512*, 239–248. [CrossRef]

11. Shang, Q.; Zhou, Y. Fabrication of transparent superhydrophobic porous silica coating for self-cleaning and anti-fogging. *Ceram. Int.* **2016**, *42*, 8706–8712. [CrossRef]

12. Liu, Y.; Zhang, K.; Yao, W.; Liu, J.; Han, Z.; Ren, L. Bioinspired structured superhydrophobic and superoleophilic stainless steel mesh for efficient oil-water separation. *Colloid Surf. A* **2016**, *500*, 54–63. [CrossRef]

13. Wu, D.; Wang, P.; Wu, P.; Yang, Q.; Liu, F.; Han, Y.; Xu, F.; Wang, L. Determination of contact angle of droplet on convex and concave spherical surfaces. *Chem. Phys.* **2015**, *457*, 63–69. [CrossRef]

14. Zhao, J.; Su, Z.; Yan, S. Thermodynamic analysis on an anisotropically superhydrophobic surface with a hierarchical structure. *Appl. Surf. Sci.* **2015**, *357*, 1625–1633. [CrossRef]

15. Luo, B.H.; Shum, P.W.; Zhou, Z.F.; Li, K.Y. Surface geometrical model modification and contact angle prediction for the laser patterned steel surface. *Surf. Coat. Technol.* **2010**, *205*, 2597–2604. [CrossRef]

16. Nickelsen, S.; Moghadam, A.D.; Ferguson, J.B.; Rohatgi, P. Modeling and experimental study of oil/water contact angle on biomimetic micro-parallel-patterned self-cleaning surfaces of selected alloys used in water industry. *Appl. Surf. Sci.* **2015**, *353*, 781–787. [CrossRef]

17. Kota, A.K.; Kwon, G.; Tuteja, A. The design and applications of superomniphobic surfaces. *NPG Asia Mater.* **2014**, *6*, e109. [CrossRef]

18. McHale, G.; Shirtcliffe, N.J.; Newton, M.I. Contact-angle hysteresis on super-hydrophobic surfaces. *Langmuir* **2004**, *20*, 10146–10149. [CrossRef]

19. Hoshian, S.; Jokinen, V.; Franssila, S. Robust hybrid elastomer/metal-oxide superhydrophobic surfaces. *Soft Matter* **2016**, *12*, 6526–6535. [CrossRef]

20. Wenzel, R.N. Resistance of solid surfaces to wetting by water. *Ind. Eng. Chem.* **1936**, *28*, 988–994. [CrossRef]

21. Yamamoto, K.; Ogata, S. 3-D thermodynamic analysis of superhydrophobic surfaces. *J. Colloid Interface Sci.* **2008**, *326*, 471–477. [CrossRef] [PubMed]

22. Nosonovsky, M.; Bhushan, B. Roughness optimization for biomimetic superhydrophobic surfaces. *Microsyst. Technol.* **2005**, *11*, 535–549. [CrossRef]

23. Bittoun, E.; Marmur, A. Optimizing super-hydrophobic surfaces: Criteria for comparison of surface topographies. *J. Adhes. Sci. Technol.* **2009**, *23*, 401–411. [CrossRef]

24. Yang, W.; Gao, H.; Zhao, Y.; Bi, K.; Li, X. Facile preparation of nitrogen-doped graphene sponge as a highly efficient oil absorption material. *Mater. Lett.* **2016**, *178*, 95–99. [CrossRef]

25. Rong, J.; Zhang, T.; Qiu, F.; Xu, J.; Zhu, Y.; Yang, D.; Dai, Y. Design and preparation of efficient, stable and superhydrophobic copper foam membrane for selective oil absorption and consecutive oil–water separation. *Mater. Des.* **2018**, *142*, 83–92. [CrossRef]

26. Liu, L.; Singh, R.; Li, G.; Xiao, G.; Webley, P.A.; Zhai, Y. Synthesis of hydrophobic zeolite X@SiO$_2$ core–shell composites. *Mater. Chem. Phys.* **2012**, *133*, 1144–1151. [CrossRef]

27. Hoshian, S.; Jokinen, V.; Somerkivi, V.; Lokanathan, A.R.; Franssila, S. Robust superhydrophobic silicon without a low surface-energy hydrophobic coating. *ACS Appl. Mater. Interfaces* **2015**, *7*, 941. [CrossRef] [PubMed]

28. Kung, C.H.; Zahiri, B.; Sow, P.K.; Mérida, W. On-demand oil-water separation via low-voltage wettability switching of core-shell structures on copper substrates. *Appl. Surf. Sci.* **2018**, *444*, 15–27. [CrossRef]

29. Cao, N.; Yang, B.; Barras, A.; Szunerits, S.; Boukherroub, R. Polyurethane sponge functionalized with superhydrophobic nanodiamond particles for efficient oil/water separation. *Chem. Eng. J.* **2017**, *307*, 319–325. [CrossRef]

30. Guselnikova, O.; Elashnikov, R.; Postnikov, P.; Svorcik, V.; Lyutakov, O. Smart, piezo-responsive polyvinylidenefluoride/polymethylmethacrylate surface with triggerable water/oil wettability and adhesion. *ACS Appl. Mater. Interfaces* **2018**, *10*, 37461–37469. [CrossRef]

31. Zhang, L.; Li, H.; Lai, X.; Su, X.; Liang, T.; Zeng, X. Thiolated graphene-based superhydrophobic sponges for oil-water separation. *Chem. Eng. J.* **2017**, *316*, 736–743. [CrossRef]

32. Hu, Y.; Zhu, Y.; Wang, H.; Wang, C.; Li, H.; Zhang, X.; Yuan, R.; Zhao, Y. Facile preparation of superhydrophobic metal foam for durable and high efficient continuous oil-water separation. *Chem. Eng. J.* **2017**, *322*, 157–166. [CrossRef]

33. Chen, X.; Liang, Y.N.; Tang, X.Z.; Shen, W.; Hu, X. Additive-free poly (vinylidene fluoride) aerogel for oil/water separation and rapid oil absorption. *Chem. Eng. J.* **2017**, *308*, 18–26. [CrossRef]

34. Yue, X.; Zhang, T.; Yang, D.; Qiu, F.; Rong, J.; Xu, J.; Fang, J. The synthesis of hierarchical porous Al$_2$O$_3$/acrylic resin composites as durable, efficient and recyclable absorbents for oil/water separation. *Chem. Eng. J.* **2017**, *309*, 522–531. [CrossRef]

35. Wang, H.; Wang, E.; Liu, Z.; Gao, D.; Yuan, R.; Sun, L.; Zhu, Y. A novel carbon nanotubes reinforced superhydrophobic and superoleophilic polyurethane sponge for selective oil-water separation through a chemical fabrication. *J. Mater. Chem. A* **2015**, *3*, 266–273. [CrossRef]

36. Zang, D.; Liu, F.; Zhang, M.; Niu, X.; Gao, Z.; Wang, C. Superhydrophobic coating on fiberglass cloth for selective removal of oil from water. *Chem. Eng. J.* **2015**, *262*, 210–216. [CrossRef]

37. Wang, Q.; Dai, B.; Bai, J.; Yang, Z.; Guo, S.; Ding, Y.; Yang, L.; Lei, P.; Han, J.; Zhu, J. Synthesis of vertically aligned composite microcone membrane filter for water/oil separation. *Mater. Des.* **2016**, *111*, 9–16. [CrossRef]

38. Cao, C.; Cheng, J. Fabrication of robust surfaces with special wettability on porous copper substrates for various oil/water separations. *Chem. Eng. J.* **2018**, *347*, 585–594. [CrossRef]

39. Qu, M.; Zhang, B.; Song, S.; Chen, L.; Zhang, J.; Cao, X. Fabrication of superhydrophobic surfaces on engineering materials by a solution-immersion process. *Adv. Funct. Mater.* **2007**, *17*, 593–596. [CrossRef]

40. Zhang, W.; Wen, X.; Yang, S. Controlled reactions on a copper surface: Synthesis and characterization of nanostructured copper compound films. *Inorg. Chem.* **2003**, *42*, 5005–5014. [CrossRef]

41. Badge, I.; Sethi, S.; Dhinojwala, A. Carbon nanotube-based robust steamphobic surfaces. *Langmuir ACS J. Surf. Colloids* **2011**, *27*, 14726–14731. [CrossRef] [PubMed]

42. Hardman, S.J.; Muhamad-Sarih, N.; Riggs, H.J.; Thompson, R.L.; Rigby, J.; Bergius, W.N.A.; Hutchings, L.R. Electrospinning superhydrophobic fibers using surface segregating end-functionalized polymer additives. *Macromolecules* **2011**, *44*, 6461–6470. [CrossRef]

43. Guo, Z.; Chen, X.; Li, J.; Liu, J.H.; Huang, X.J. ZnO/CuO hetero-hierarchical nanotrees array: Hydrothermal preparation and self-cleaning properties. *Langmuir* **2011**, *27*, 6193–6200. [CrossRef] [PubMed]

44. Wu, R.; Liang, S.; Pan, A.; Yuan, Z.; Tang, Y.; Tan, X.P.; Guan, D.K.; Yu, Y. Fabrication of nano-structured super-hydrophobic film on aluminum by controllable immersing method. *Appl. Surf. Sci.* **2012**, *258*, 5933–5937. [CrossRef]

45. Pan, L.; Dong, H.; Bi, P. Facile preparation of superhydrophobic copper surface by HNO$_3$ etching technique with the assistance of CTAB and ultrasonication. *Appl. Surf. Sci.* **2010**, *257*, 1707–1711. [CrossRef]

46. Khosravi, M.; Azizian, S. Preparation of superhydrophobic and superoleophilic nanostructured layer on steel mesh for oil-water separation. *Sep. Purif. Technol.* **2017**, *172*, 366–373. [CrossRef]

47. Heni, W.; Vonna, L.; Fioux, P.; Vidal, L.; Haidara, H. Ultrasonic cavitation test applied to thin metallic films for assessing their adhesion with mercaptosilanes and surface roughness. *J. Mater. Sci.* **2014**, *49*, 6750–6761. [CrossRef]

48. Ma, F.M.; Hao, Q.Y.; Zhang, Y.; Ruan, M.; Yu, Z.L. Fabrication of copper-based superhydrophobic surface by redox etching. *Sci. Technol. Eng.* **2013**, *14*, 026.

49. Zhang, C.L.; Zhang, F.; Song, L.; Zeng, R.C.; Li, S.Q.; Han, E.H. Corrosion resistance of a superhydrophobic surface on micro-arc oxidation coated Mg-Li-Ca alloy. *J. Alloys Compd.* **2017**, *728*, 815–826. [CrossRef]

50. Ma, Q.; Cheng, H.; Fane, A.G.; Wang, R.; Zhang, H. Recent development of advanced materials with special wettability for selective oil/water separation. *Small* **2016**, *12*, 2186–2202. [CrossRef]

51. Liu, Y.Q.; Zhang, Y.L.; Fu, X.Y.; Sun, H.B. Bioinspired underwater superoleophobic membrane based on a graphene oxide coated wire mesh for efficient oil/water separation. *ACS Appl. Mater. Interfaces* **2015**, *7*, 20930–20936. [CrossRef] [PubMed]

52. Lin, X.; Choi, M.; Heo, J.; Jeong, H.; Park, S.; Hong, J. Cobweb-inspired superhydrophobic multi-scaled gating membrane with embedded network structure for robust water-in-oil emulsion separation. *ACS Sustain. Chem. Eng.* **2017**, *5*, 3448–3455. [CrossRef]
53. Cao, H.; Fu, J.; Liu, Y.; Chen, S. Facile design of superhydrophobic and superoleophilic copper mesh assisted by candle soot for oil water separation. *Colloid Surf. A* **2017**, *537*, 294–302. [CrossRef]
54. Zhang, D.; Li, L.; Wu, Y.; Sun, W.; Wang, J.; Sun, H. One-step method for fabrication of superhydrophobic and superoleophilic surface for water-oil separation. *Colloid Surf. A* **2018**, *552*, 32–38. [CrossRef]
55. Yuan, Z.; Wang, X.; Bin, J.; Peng, C.; Xing, S.; Wang, M.; Xiao, J.; Zeng, J.; Xie, Y.; Xiao, X. A novel fabrication of a superhydrophobic surface with highly similar hierarchical structure of the lotus leaf on a copper sheet. *Appl. Surf. Sci.* **2013**, *285*, 205–210. [CrossRef]
56. Li, J.; Huang, R.H.; Zhou, P.; Xu, Y.; Li, Z.P.; Wang, L.F. Influencing factors of splashing caused by engine oil droplets impacting on wall in the crankcase. *Trans. CSICE* **2017**, *6*, 548–553. (In Chinese)
57. Erbil, H.Y.; Demirel, A.L.; Avci, Y.; Mert, O. Transformation of a simple plastic into a superhydrophobic surface. *Science* **2003**, *299*, 1377–1380. [CrossRef]
58. Lafuma, A.; Quéré, D. Superhydrophobic states. *Nat. Mater.* **2003**, *2*, 457–460. [CrossRef]
59. Barthlott, W.; Neinhuis, C. Purity of the sacred lotus, or escape from contamination in biological surfaces. *Planta* **1997**, *202*, 1–8. [CrossRef]
60. Liu, K.; Jiang, L. Bio-inspired design of multiscale structures for function integration. *Nano Today* **2011**, *6*, 155–175. [CrossRef]
61. Cao, H.; Gu, W.; Fu, J.; Liu, Y.; Chen, S. Preparation of superhydrophobic/oleophilic copper mesh for oil-water separation. *Appl. Surf. Sci.* **2017**, *412*, 599–605. [CrossRef]
62. Jiang, Z.X.; Geng, L.; Huang, Y.D. Design and fabrication of hydrophobic copper mesh with striking loading capacity and pressure resistance. *J. Phys. Chem. C* **2010**, *114*, 9370–9378. [CrossRef]

Article

Preparation of Superhydrophobic Steel Surfaces with Chemical Stability and Corrosion

Chongwei Du [1,2], Xiaoyan He [1,2], Feng Tian [1,2], Xiuqin Bai [1,2,*] and Chengqing Yuan [1,2]

[1] Reliability Engineering Institute, National Engineering Research Center for Water Transport Safety, Wuhan University of Technology, Wuhan 430063, China; cwd@whut.edu.cn (C.D.); hexiaoyan@whut.edu.cn (X.H.); tianfeng221@whut.edu.cn (F.T.); ycq@whut.edu.cn (C.Y.)

[2] Key Laboratory of Marine Power Engineering and Technology, Ministry of Transport, Wuhan University of Technology, Wuhan 430063, China

* Correspondence: xqbai@whut.edu.cn

Received: 25 May 2019; Accepted: 18 June 2019; Published: 20 June 2019

Abstract: Corrosion seriously limits the long-term application of Q235 carbon steel. Herein, a simple fabrication method was used to fabricate superhydrophobic surfaces on Q235 carbon steel for anticorrosion application. The combination of structure and the grafted low-surface-energy material contributed to the formation of superhydrophobic steel surfaces, which exhibited a water contact angle of 161.6° and a contact angle hysteresis of 0.8°. Meanwhile, the as-prepared superhydrophobic surface showed repellent toward different solutions with pH ranging from 1 to 14, presenting excellent chemical stability. Moreover, the acid corrosive liquid (HCl solution with pH of 1) maintained sphere-like shape on the as-prepared superhydrophobic surface at room temperature, indicating superior corrosion resistance. This work provides a simple method to fabricate superhydrophobic steel surfaces with chemical stability and corrosion resistance.

Keywords: carbon steel; chemical etching; superhydrophobic; chemical stability; corrosion resistance

---

1. Introduction

Q235 carbon steel, medium-low strength steel with low-cost, has a large number of applications in industrial fields, especially in marine environment [1]. However, Q235 carbon steel is prone to corrosion, especially in a neutral or acidic environment. The natural environment, including the atmosphere, water, and soil, is usually neutral or weakly acidic to induce serious corrosion of Q235 carbon steel [2,3]. In order to inhibit corrosion of Q235, various methods have been applied for improving anti-corrosion performance of Q235. Among them, cathodic protection and anti-corrosion coatings are widely used to prevent corrosion. However, the high cost and the consumption of non-ferrous metals limited the applications of cathodic protection technology. Anti-corrosion coatings could effectively isolate or inhibit the direct contact between the protected metals and the corrosive media by their shielding and resistance effect to slow the corrosion rate of the protected metal [4]. At present, there were various methods used to prepare anti-corrosion surfaces on metal substrates, including chemical etching [5], electrochemical deposition [6], anodizing [7], sol-gel [8], thermal oxidation [9], laser structuration [10], and thermal spraying [11]. The superhydrophobic surfaces have a very important position as a kind of anti-corrosion coating. The strong repellent ability against water of the superhydrophobic surface can greatly reduce the contact area of corrosive liquid and the surface, and hinder the migration of corrosive ions to metal materials, leading to a decrease in the corrosion rate of materials [12]. Chen et al. fabricated a superhydrophobic aluminum surface on steel substrate by suspension flame spraying in large-scale fabrication [13]. The superhydrophobic aluminum surface showed excellent anticorrosive performance with a contact angle of 151°. Xiang et al. prepared a superhydrophobic surface on mild steel. The fabrication process involved two steps, including formation of porous ZnO coatings and modification of myristic acid [14]. These studies were based on coatings to achieve superhydrophobicity. However, adhesion strength between the coatings and the substrate was a big problem which needs to be solved. Kietzig et al. fabricated dual scale roughness structures on different metals by femtosecond laser irradiation to achieve a hydrophobic surface with a contact angle above 150° [10]. Jagdheesh et al. used a nanosecond laser light source on the stainless-steel plate to produce a single roughness micropattern. The as-prepared surface showed excellent superhydrophobicity with a roll-off angle of 5° and a static contact angle of 180° [15]. However, this method needs special equipment.

In our study, superhydrophobic surface on Q235 carbon steel was prepared by a two-step process, involving the formation of an irregular needle-like structure and fluorination treatment. The morphology of the needle-like microstructure was in-situ formed by chemical etching on carbon steel. The combination of structured surface and low surface material contributed to the formation of the superhydrophobic surface. The anti-wetting performance of the as-prepared superhydrophobic surface was further studied. The results demonstrated that the superhydrophobic surface exhibited a high water-contact angle (~161.6°) and a low contact angle hysteresis (~0.8°), suggesting the Cassie–Baxter state on the surface. Meanwhile, the bounce phenomenon of water droplets on the surface was photographed. More importantly, the as-prepared samples could maintain good hydrophobic properties toward the acid solution and base solution with pHs ranging from 1 to 14. The corrosion resistance to acid solution of the as-prepared superhydrophobic surface was also examined. The inspiring results might provide insight into developing superhydrophobic surfaces on iron substrates for enhanced performances.

2. Materials and Methods

2.1. Materials

The commercially available Q235 carbon steel (GB, containing element content: Fe ≥ 98.675%, C 0.14%~0.22%, Mn 0.30~0.65%, Si ≤ 0.210%, S ≤ 0.200%, P ≤ 0.045%) with a diameter of 10 mm and a thickness of 3 mm were used. (1H,1H,2H,2H-perfluorodecyl)-triethoxysilane (PFTEOS, 97%) was purchased from Aladdin Industrial Corporation (Shanghai, China). All chemicals were used as received without any further purification.

2.2. Sample Preparation

Q235 carbon steel discs were polished and washed by ethanol and deionized water, subsequently drying with flowing nitrogen. Q235 carbon steel discs were placed in Piranha solution at 35 °C for 15 min and then washed by deionized water for three time. Piranha solution was prepared by mixing concentrated sulfuric acid (95.0%~98.0%, AR) and hydrogen peroxide (30%, AR) in a volume ratio of 7:3. Finally, the treated steels were soaked in the 1% (v/v) ethanol solution of PFTEOS at 35 °C for 24 h, and then drying with nitrogen.

2.3. Characterization

To demonstrate the specific microstructure of the sample surface, a preliminary observation of the sample surface was performed using a field scanning electron microscope (FESEM, Zeiss Ultra Plus, Heidenheim, Germany). The roughness of samples was tested by a contourgraph (Huazhong University of Science and Technology, Wuhan, China). Analysis of the elemental composition of the sample surface was using energy-dispersive spectroscopy (EDS, Oxford, Abingdon, UK) at an accelerating voltage of 8 kV. Further analysis of the chemical composition of the sample was by X-ray photoelectron spectroscopy (XPS, ESCALAB 250Xi, Waltham, MA, USA) using Al-Kα as the radiation source. The contact angle of the samples and controls were measured by a contact angle meter (Dataphysics OCA 15EC, Stuttgart, Germany). The contact angle of the sample toward acid solutions and base solutions was measured using a standard solution of pH ranging from 1 to 14 to evaluate the surface anti-wetting properties of the sample under acidic and alkaline conditions. The droplet size used to measure the water contact angle was 3 µL. Meanwhile, the bounce phenomenon of the droplet on the surface of the sample was observed and photographed with the droplet size of 5 µL. The contact angle hysteresis was calculated by adding and shrinking a small drop of deionized water (drop volume, 9 µL) on the surface. The image analysis system was used to digitally record and measure the advancing contact angle and the receding contact angle. The difference between advancing contact angle and the receding contact angle determined water contact angle hysteresis.

Electrochemical measurements were obtained in 3.5 wt % NaCl solution by a CS300 electrochemical workstation with a standard three-electrode cell (CorrTest Instrument Co., Ltd., Wuhan, China). The saturated calomel electrode was used as the reference electrode. The platinum foil of 10×10 mm^2 surface area was used as a counter electrode. The working electrode with a circular exposed area of 0.785 cm^2. Electrochemical impedance spectra were carried out with signals of 10 mV amplitude in the frequency spectrum ranging from 100 kHz to 0.01 Hz. Potentiodynamic polarization curves were acquired with the potential range of −200 mV to 200 mV versus E_{ocp} at a scan rate of 1 mV/s. The corrosion potential (E_{corr}) and the corrosion current density (I_{corr}) were calculated by CView 2.6.

3. Results and Discussion

Figure 1 showed the morphological features of the polished Q235 carbon steel and the structured Q235 carbon steel. The polished carbon steel was very smooth with an average roughness (R_a) of 39 nm (Figure 1a). The polished Q235 carbon steel substrates were immersed into the Piranha solution to obtain the structured surface. The addition of H_2O_2 in the acid solutions could accelerate the etching process and oxidation reactions [16]. The etching process of H_2O_2 and H_2SO_4 resulted in the microstructures on the Q235 carbon steel. The structured Q235 carbon steel exhibited a rough surface with complex and regular needle-like microstructures (Figure 1(b-1)). The needle-like microstructures were about 200 nm in length and 10 nm in width, resulting in nano-pores with diameter of 100 nm. The roughness significantly increased to 491 nm. The further grafting of PFTEOS on the structured Q235 carbon steel surface slightly reduced the roughness to 423 nm. Moreover, the topography of the structured surface after modification of PFTEOS was similar with that of structured Q235 carbon steel surface (Figure 1c). The microstructures provided enough space to entrap air layer, which was an important factor to construct superhydrophobic surface.

The other factor that contributes to the surface wetting properties is the chemical compositions of the surface. The EDS results of polished Q235 carbon steel surface, structured Q235 carbon steel surface and structured surface after modification of PFTEOS were showed in Figure 2. The polished Q235 carbon steel surface only contained element Fe (Figure 2a). The structured Q235 carbon steel surface contained element Fe, O and C (Figure 2b). It was obvious that the structured steel surface after modified PFTEOS consisted of elements Fe, F, and Si (Figure 2c), indicating that PFTEOS has been successfully grafted on the surface of the sample.

To further analyze the composition of structured surfaces before/after modified PFTEOS, XPS analyses of the surfaces were employed (Figure 3). The elemental compositions of the structured Q235 carbon steel surface was Fe, O, C (Figure 3(a-1)). While, the extra elements F and Si appeared on the structured steel surface after modification of PFTEOS, indicating that PFTEOS were successfully attached to the substrate surface after modification (Figure 3(b-1)). The O 1s spectra of the structured Q235 carbon steel surface resolved into two components, namely C–O (530.98 eV) and FeOOH (529.68 eV) (Figure 3(a-2)), indicating that FeOOH can be formed on the surface of Q235 carbon steel after the chemically etched by Piranha solution. The C 1s spectrum acquired from the structured Q235 carbon steel surface, was resolved into three components, namely C–C/C–H (284.60 eV), C–O (286.10 eV) and C=O (288.22 eV) [17] (Figure 3(a-3)). However, the spectra of the structured surface after modification of PFTEOS were different. The O 1s spectra of the structured steel surface after modification of PFTEOS resolved into C=O (532.60 eV) and Fe–O–Si (530.75 eV) (Figure 3(b-2)). The C 1s spectrum was resolved into three components, namely C–C/C–H (284.94 eV), C–O (286.01 eV) and C=O (288.24 eV), CF$_2$ (291.62 eV) and CF$_3$ (293.82 eV) [18] (Figure 3(b-3)). The F 1s proved the existence of C–F (689.02 eV) (Figure 3(b-4)). The XPS spectrum of Si 1s showed the formation of Si–O (103.09 eV) (Figure 3(b-5)). The O 1s peak of 530.75 eV of the modified Q235 carbon steel surface indicated the formation of Fe–O–Si bond by the chemical reaction of the organosilane and the hydroxylated aluminum surface (Figure 3(b-2)). The hydrolysis reaction between the silicon ethyoxyl (Si–OCH$_2$CH$_3$) of PFTEOS and the hydroxyl group in FeOOH on the surface was to form a self-assembled film by stable chemical binding. These results indicated that the PFTEOS was successfully chemical bonding to the carbon steel surface. As a result, the CF$_3$ and CF$_2$ groups were rich in the surface, which could improve the hydrophobic properties of Q235 carbon steel.

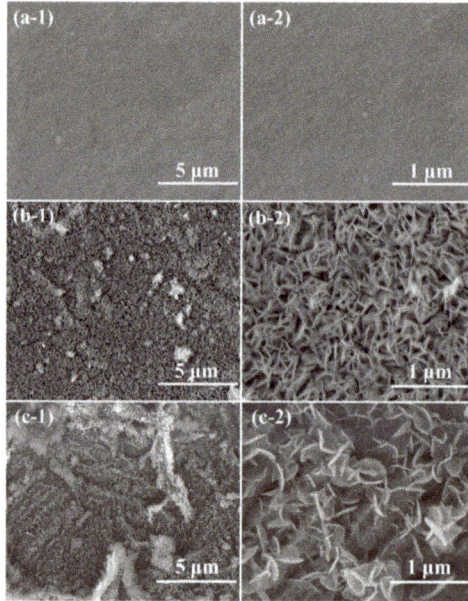

Figure 1. SEM images of polished Q235 carbon steel surface (**a**), structured Q235 carbon steel surface (**b**) and structured surface after modification of PFTEOS (**c**). (-2 is the high magnification image of -1, respectively).

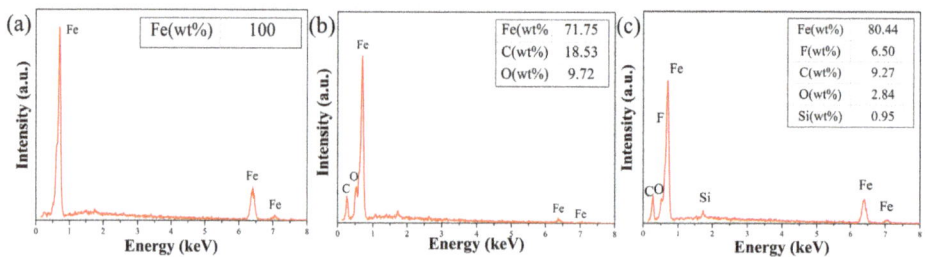

Figure 2. EDS of polished Q235 carbon steel surface (**a**), structured Q235 carbon steel surface (**b**), and structured surface after modification of PFTEOS (**c**).

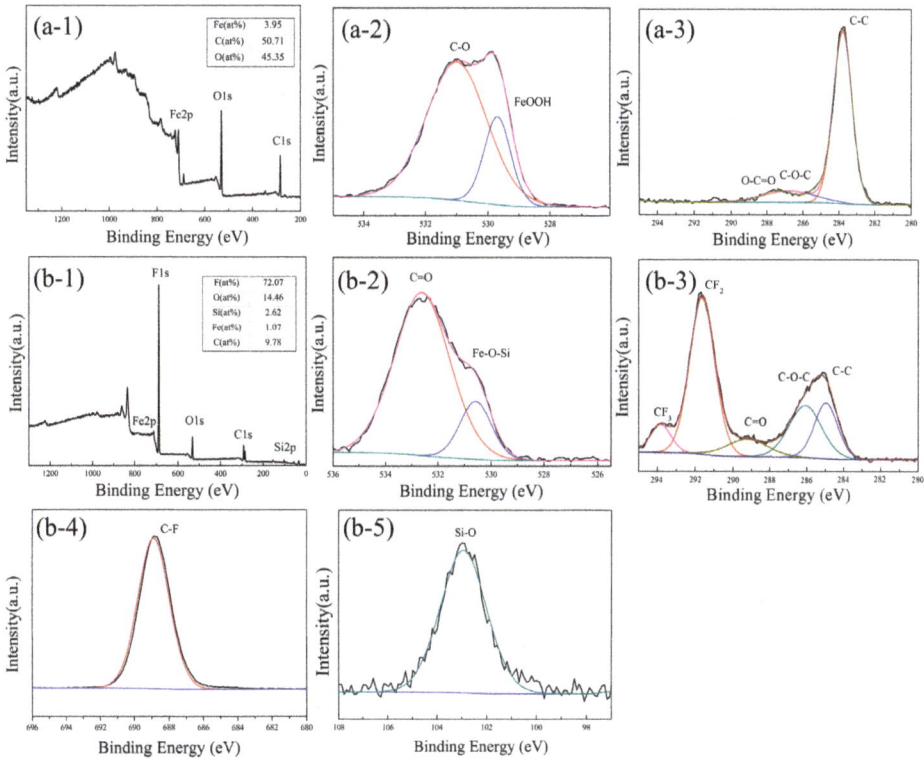

Figure 3. XPS spectra detected from structured Q235 carbon steel surface: the XPS survey spectra (**a-1**), O 1s spectra (**a-2**) and C 1s spectra (**a-3**) and structured surface after modification of PFTEOS: the XPS survey spectra (**b-1**), O 1s spectra (**b-2**), C 1s spectra (**b-3**), F 1s spectra (**b-4**) and Si 2p spectra (**b-5**). All spectra were corrected to the polluted C 1s spectra.

Water contact angle (WCA) and contact angle hysteresis (CAH) were utilized to characterize the hydrophobic properties of the sample surfaces. The water contact angles of polished steel surface and the polished steel surface with grafted PFTEOS were 90.3° and 99.1°, respectively (Figure 4a,b). WCA and CAH before and after PFTEOS modification were shown in Figure 4c,d. It could be clearly seen that the hydrophobic properties of the structured steel surface after modification of PFTEOS are much better than those of the structured Q235 carbon steel (Figure 4). The structured sample modified by PFTEOS has a WCA of 161.6° ± 0.5° and a CAH of 0.8° ± 0.5° (Figure 4d), which was a superhydrophobic surface. High WCA and low CAH indicated that the modified surface was in the Cassie–Baxter state. The surface of the structured surface is a rough surface with microstructure (Figure 4d), which provided enough space for the solid surface to lock the air and form an air layer. This air layer can effectively isolate the direct contact between the solid surface and the liquid to achieve hydrophobicity. According to the formula given by the Cassie–Baxter model definition [19], the percentage of solid surface area and gas-liquid area on the surface of the object can be calculated.

$$\cos \theta_r = f_1 \cos \theta - f_2$$

θ_r: the contact angle on the structured surfaces after modification.

θ: the contact angle on the smooth surface after modification.

f_1: The surface area of the liquid in contact with solid divided by the projected area.

f_2: The surface area of the liquid in contact with air divided by the projected area.

   Water contact angle of polished surface after modification is 99.1° ± 0.5° and the water contact angle of the structured surface after modification is 161.6° ± 0.5°. It is known that $f_1 + f_2 = 1$. After calculation, f_1 is about 6.1%, and f_2 is about 93.9%. The contact area between the air and the liquid on the surface of the modified sample is much larger than the contact area between the solid and the liquid, which suggested that the superhydrophobic surface has been achieved. Therefore, the microstructure and the modification of PFTEOS contribute to the superhydrophobicity of Q235 carbon steel, resulting in an air layer to reduce the solid-liquid contact area.

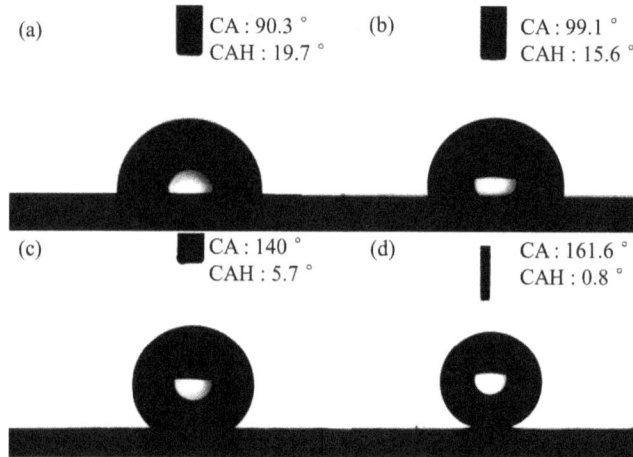

Figure 4. Contact angle of water on the polished Q235 carbon steel surface (**a**), polished steel surface after modification of PFTEOS (**b**), microstructured steel surface (**c**), and structured steel surface after modification of PFTEOS (**d**).

   The bounce phenomenon of superhydrophobic Q235 carbon steel surface was showed in Figure 5. It took about 560 ms for the droplet to stay on the surface. The water droplet would compress, retract, and bounce off superhydrophobic Q235 carbon steel surface and the bounce phenomenon was repeated three times. As a result, the special wettability was highly similar with the surface of lotus leaf. This phenomenon further indicated that the superhydrophobic Q235 carbon steel surface was in Cassie–Baxter state with the extremely low water affinity.

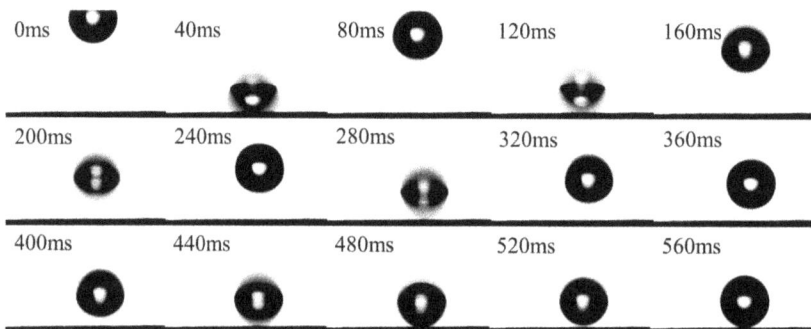

Figure 5. Bounce phenomenon of superhydrophobic Q235 carbon steel surface (5 µL).

The stability of superhydrophobic surfaces over a wide pH-range is significant for practical applications in marine environment. Particularly, media with different pH values will impact performance of superhydrophobic surfaces. It was excited to note that the as-prepared superhydrophobic surfaces on Q235 carbon steel still maintained high hydrophobicity toward both acid and alkaline solutions (Figure 6), suggesting excellent chemical stability. However, the structured Q235 carbon steel could not maintain hydrophobicity toward acid solutions. The contact angles of both the structured Q235 carbon steel and the superhydrophobic Q235 carbon steel were stable when they were tested by the alkaline solutions. As we known, iron does not readily react with hydroxide ions at room temperature, so alkaline solutions had no influence on their wettability. The acid solutions significantly influenced the wettability of the structured Q235 carbon steel with an obvious decrease to 74.2°. While, the as-prepared superhydrophobic surfaces on Q235 carbon steel still maintained high hydrophobicity toward acid solution with a pH of 1.

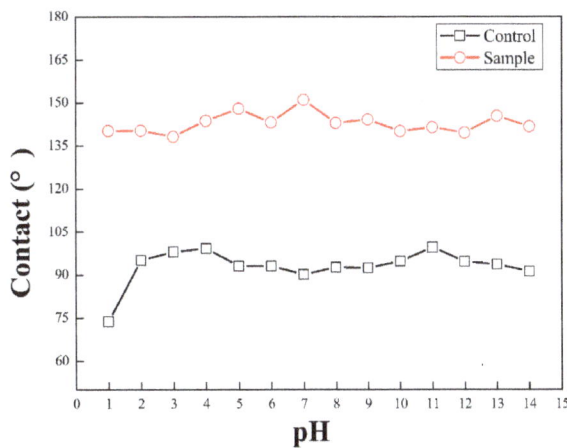

Figure 6. Comparison of different pH environments contact angle of structured Q235 carbon steel and superhydrophobic Q235 carbon steel.

To evaluate the corrosion resistance of the superhydrophobic surface, the EIS and potentiodynamic polarization curves were performed. EIS analyses were used to assess their corrosion resistances and the Nyquist plots were shown in Figure 7a. The impedance of superhydrophobic Q235 carbon steel is much higher than those of the untreated Q235 carbon steel and structured Q235 carbon steel, indicating that the corrosion resistance of superhydrophobic samples is better than the other samples [20]. The potentiodynamic polarization results were consistent with the results obtained from EIS. The potentiodynamic polarization results were shown in Figure 7b. In the polarization curves, the lower I_{corr} and higher E_{corr} indicates the higher corrosion resistance [21]. The I_{corr} of the superhydrophobic Q235 carbon steel is about 0.577 $\mu A/cm^2$, which is much lower than those of the untreated Q235 carbon steel (6.601 $\mu A/cm^2$) and structured Q235 carbon steel (3.917 $\mu A/cm^2$). Meanwhile, the E_{corr} of the superhydrophobic Q235 carbon steel is positive-shifting comparing with the other samples. The results suggest that the superhydrophobic Q235 carbon steel samples exhibit excellent corrosion resistance.

To visually exhibit the superhydrophobicity and anticorrosion performance of the superhydrophobic Q235 carbon steel, a highly corrosive liquid droplet (pH = 1 HCl + $CuSO_4$) was dropped onto the untreated Q235 carbon steel, structured Q235 carbon steel, superhydrophobic Q235 carbon steel (Figure 8). It could be seen that the untreated Q235 carbon steel began to corrode at 10 s and the corrosion area gradually increased in the 60 s (Figure 8a). The structured samples had a certain corrosion resistance. The damage of the structured Q235 carbon steel occurred after contact with the corrosive liquid droplet for 7 min. The ball-like shape of the droplets was not be maintained, and the color of the corrosive liquid droplets turned to reddish brown, indicating the formation of ferric ions (Figure 8b). The oxidation film on the structured Q235 carbon steel caused by the H_2O_2 in H_2SO_4 solution could have contributed to the corrosion resistance [22]. Therefore, the structured Q235 carbon steel had better anticorrosion performance than the untreated Q235 carbon steel. However, the droplets of the highly-corrosive liquid solution on the superhydrophobic Q235 carbon steel retained their sphere-like shape for at least 10 min (Figure 8c). The results demonstrated that the superhydrophobic Q235 carbon steel was greatly stable toward acid solutions. The air layer captured on the superhydrophobic Q235 carbon steel could isolate the corrosive liquid droplet and the substrate to obtain corrosion resistance.

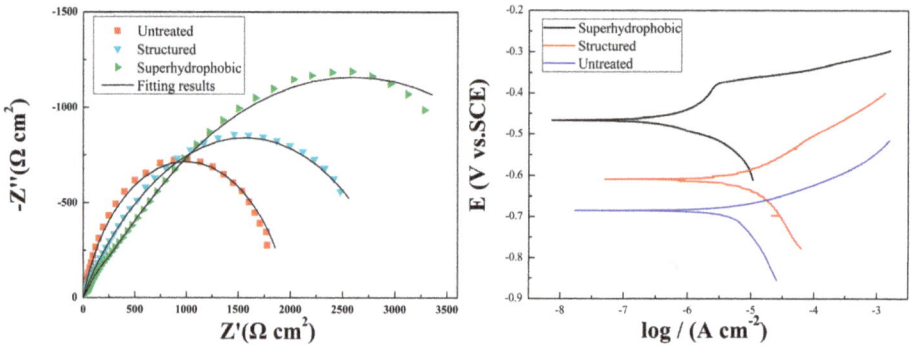

Figure 7. Nyquist plots (**a**) and Potentiodynamic polarization curves (**b**) of untreated Q235 carbon steel samples, structured Q235 carbon steel samples, and superhydrophobic Q235 carbon steel samples.

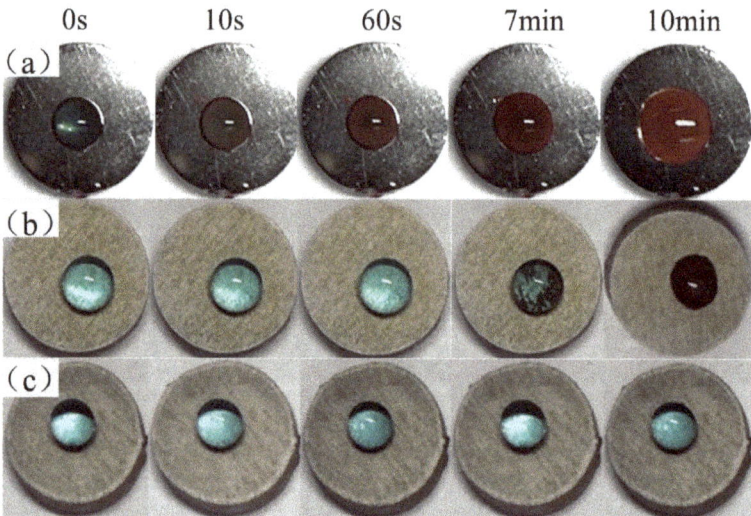

Figure 8. Sequential images of a highly corrosive liquid droplet (pH = 1 HCl + $CuSO_4$) on the surfaces of untreated Q235 carbon steel (**a**), structured Q235 carbon steel (**b**), superhydrophobic Q235 carbon steel (**c**).

4. Conclusions

Our study proposed a simple and efficient to fabricate superhydrophobic surface on Q235 carbon steel. The surface of Q235 carbon steel was chemically etched by Piranha solution to produce FeOOH with microstructure and then PFTEOS could graft onto the surface by covalent bonding with FeOOH.

The microstructure and modification of PFTEOS resulted in superhydrophobic surface on Q235 carbon steel. The as-prepared surface could show repellent toward water, acid solutions, and alkaline solutions. Meanwhile, the superhydrophobic surface on Q235 carbon steel was in Cassie–Baxter state, which the existed air layer could isolated solid-liquid interface to inhibit corrosion. This provides a reference for future research on steel corrosion protection.

Author Contributions: C.D. carried out the literature search, figures, data collection, data analysis, data interpretation and writing. X.H., F.T. and X.B. conceived for study design. X.H. and F.T. participated in data collection. X.H., F.T., X.B. and C.Y. contributed to data analysis. X.H., X.B. and C.Y. were responsible for writing and revision.

Funding: This work was supported by China Postdoctoral Science Foundation (Grant No. 2017M622542), and Hubei Postdoctoral Sustentation Foundation, China (Grant No. G22).

Acknowledgments: The authors would like to thank the supporting by the Material research and testing center (Wuhan university of technology).

Conflicts of Interest: The authors declare no conflict of interest.

References

1. Usher, K.M.; Kaksonen, A.H.; Cole, I.; Marney, D. Critical review: Microbially influenced corrosion of buried carbon steel pipes. *Int. Biodeterior. Biodegrad.* **2014**, *93*, 84–106. [CrossRef]
2. Strickland, D.M. The resistivity of iron and its application to the chemical industry. *Ind. Eng. Chem.* **1923**, *15*, 566–569. [CrossRef]
3. Guo, W.; Chen, S.; Feng, Y.; Yang, C. Investigations of triphenyl phosphate and Bis-(2-ethylhexyl) phosphate self-assembled films on iron surface using electrochemical methods, fourier transform infrared spectroscopy, and molecular simulations. *J. Phys. Chem. C* **2007**, *111*, 3109–3115. [CrossRef]
4. Pu, N.W.; Shi, G.N.; Liu, Y.M.; Sun, X.; Chang, J.K.; Sun, C.L.; Ger, M.-D.; Chen, C.Y.; Wang, P.C.; Peng, Y.Y.; et al. Graphene grown on stainless steel as a high-performance and ecofriendly anti-corrosion coating for polymer electrolyte membrane fuel cell bipolar plates. *J. Power Sources* **2015**, *282*, 248–256. [CrossRef]
5. Huang, Y.; Sarkar, D.; Chen, X.-G. Fabrication of corrosion resistance micro-nanostructured superhydrophobic anodized aluminum in a one-step electrodeposition process. *Metals* **2016**, *6*, 47. [CrossRef]
6. Zhao, G.; Xue, Y.; Huang, Y.; Ye, Y.; Walsh, F.C.; Chen, J.; Wang, S. One-step electrodeposition of a self-cleaning and corrosion resistant Ni/WS$_2$ superhydrophobic surface. *RSC Adv.* **2016**, *6*, 59104–59112. [CrossRef]
7. Liang, J.; Liu, K.; Wang, D.; Li, H.; Li, P.; Li, S.; Su, S.; Xu, S.; Luo, Y. Facile fabrication of superhydrophilic/superhydrophobic surface on titanium substrate by single-step anodization and fluorination. *Appl. Surf. Sci.* **2015**, *338*, 126–136.
8. Yang, H.; Pi, P.; Cai, Z.Q.; Wen, X.; Wang, X.; Cheng, J.; Yang, Z.R. Facile preparation of super-hydrophobic and super-oleophilic silica film on stainless steel mesh via sol-gel process. *Appl. Surf. Sci.* **2010**, *256*, 4095–4102. [CrossRef]
9. Zhang, F.; Zhang, C.; Song, L.; Zeng, R.; Li, S.; Cui, H. Fabrication of the superhydrophobic surface on magnesium alloy and its corrosion resistance. *J. Mater. Sci. Technol.* **2015**, *31*, 1139–1143. [CrossRef]
10. Kietzig, A.M.; Hatzikiriakos, S.G.; Englezos, P. Patterned superhydrophobic metallic surfaces. *Langmuir* **2009**, *25*, 4821–4827. [CrossRef]
11. Men, X.; Zhang, Z.; Yang, J.; Zhu, X.; Wang, K.; Jiang, W. Spray-coated superhydrophobic coatings with regenerability. *New J. Chem.* **2011**, *35*, 881–886. [CrossRef]
12. He, G.; Lu, S.; Xu, W.; Yu, J.; Wu, B.; Cui, S. Fabrication of durable superhydrophobic electrodeposited tin surfaces with tremella-like structure on copper substrate. *Surf. Coatings Technol.* **2017**, *309*, 590–599. [CrossRef]
13. Chen, X.; Yuan, J.; Huang, J.; Ren, K.; Liu, Y.; Lu, S.; Li, H. Large-scale fabrication of superhydrophobic polyurethane/nano-Al$_2$O$_3$ coatings by suspension flame spraying for anti-corrosion applications. *Appl. Surf. Sci.* **2014**, *311*, 864–869. [CrossRef]
14. Xiang, T.; Han, Y.; Guo, Z.; Wang, R.; Zheng, S.; Li, S.; Li, C.; Dai, X. Fabrication of inherent anticorrosion superhydrophobic surfaces on metals. *ACS Sustain. Chem. Eng.* **2018**, *6*, 5598–5606. [CrossRef]

15. Jagdheesh, R.; Diaz, M.; Marimuthu, S.; Ocana, J.L. Robust fabrication of μ-patterns with tunable and durable wetting properties: hydrophilic to ultrahydrophobic via a vacuum process. *J. Mater. Chem. A* **2017**, *5*, 7125–7136. [CrossRef]

16. Tsujino, K.; Matsumura, M. Morphology of nanoholes formed in silicon by wet etching in solutions containing HF and H_2O_2 at different concentrations using silver nanoparticles as catalysts. *Electrochim. Acta* **2007**, *53*, 28–34. [CrossRef]

17. Olivares, O.; Likhanova, N.V.; Gómez, B.; Navarrete, J.; Llanos-Serrano, M.E.; Arce, E.; Hallen, J.M. Electrochemical and XPS studies of decylamides of α-amino acids adsorption on carbon steel in acidic environment. *Appl. Surf. Sci.* **2006**, *252*, 2894–2909. [CrossRef]

18. Vandencasteele, N.; Reniers, F. Plasma-modified polymer surfaces: Characterization using XPS. *J. Electron Spectros. Relat. Phenomena* **2010**, *178–179*, 394–408. [CrossRef]

19. Parry, V.; Berthomé, G.; Joud, J.C. Wetting properties of gas diffusion layers: Application of the Cassie–Baxter and Wenzel equations. *Appl. Surf. Sci.* **2012**, *258*, 5619–5627. [CrossRef]

20. Xue, Y.; Wang, S.; Zhao, G.; Taleb, A.; Jin, Y. Fabrication of Ni–Co coating by electrochemical deposition with high super-hydrophobic properties for corrosion protection. *Surf. Coatings Technol.* **2019**, *363*, 352–361. [CrossRef]

21. Chellouli, M.; Chebabe, D.; Dermaj, A.; Erramli, H.; Bettach, N.; Hajjaji, N.; Casaletto, M.P.; Cirrincione, C.; Privitera, A.; Srhiri, A. Corrosion inhibition of iron in acidic solution by a green formulation derived from *Nigella sativa* L. *Electrochim. Acta* **2016**, *204*, 50–59. [CrossRef]

22. Homjabok, W.; Permpoon, S.; Lothongkum, G. Pickling behavior of AISI 304 stainless steel in sulfuric and hydrochloric acid solutions. *J. Met. Mater. Miner.* **2017**, *20*, 1–6.

coatings

MDPI

Article

Super-Hydrophobic Co–Ni Coating with High Abrasion Resistance Prepared by Electrodeposition

Yanpeng Xue [1,*,†], Shuqiang Wang [1,†], Peng Bi [1], Guochen Zhao [2] and Ying Jin [1,*]

[1] National Center for Materials Service Safety, University of Science and Technology Beijing, Beijing 100083, China; s20171185@xs.ustb.edu.cn (S.W.); s20161187@ustb.cn (P.B.)
[2] Shandong Provincial Key Laboratory for High Strength Lightweight Metallic Materials, Advanced Materials Institute, Qilu University of Technology (Shandong Academy of Sciences), Jinan 250000, China; zhaogch@sdas.org
*   Correspondence: yanpengxue@ustb.edu.cn (Y.X.); yjin@ustb.edu.cn (Y.J.)
†   These authors contributed equally to this work.

Received: 19 February 2019; Accepted: 29 March 2019; Published: 2 April 2019

Abstract: Although super-hydrophobic surfaces have great application prospects in industry, their preparation cost and mechanical durability have limited their practical utilization. In this work, we presented a new low-cost process preparation for super-hydrophobic Co–Ni coating on carbon steel substrate via an electrodeposition route. The deposited Co–Ni coating with cauliflower-shaped micro-nano structures exhibited high super-hydrophobic properties with water contact angles over 161° after modification with 1H,1H,2H,2H-Perfluorooctyltrichlorosilane (PFTEOS). Evaluated by the linear abrasion methods, the super-hydrophobic coating can maintain super-hydrophobicity after abrasion distance of 12 m under the applied pressure of 5 kPa, which was attributed to the high cobalt content of the Co–Ni coating. Moreover, electrochemical tests showed that the super-hydrophobic Co–Ni coatings exhibited a good anti-corrosion performance thus providing an adequate protection to the carbon steel substrates.

Keywords: electrochemical deposition; Co–Ni coating; super-hydrophobic surface; mechanical durability; corrosion protection

1. Introduction

Biomimetic materials with hierarchical structures, which are compatible with environmental requirements, have become a major concern in the scientific community recently [1,2]. It is well known that many naturally occurring surfaces, including lotus leaves and water striders, possess super-hydrophobic and self-cleaning properties, having water contact angles (WCA) greater than 150° and a small contact angle hysteresis of less than 10° [3,4]. Super-hydrophobic coatings have received extensive research attentions due to their potential applications in corrosion protection [5], anti-icing [6], water/oil separation [7] and drag reduction [8]. Generally speaking, the wetting property of a solid surface is determined by the combination of chemical compositions and microstructures. To fabricate the super-hydrophobic surface, a large number of techniques have been developed, such as etching [9], femtosecond laser machining [10], chemical vapor deposition [11], anodic oxidation [12] and colloidal coating [13]. Certain fabrication techniques to develop super-hydrophobic surfaces have been limited in industrial applications because they are relatively expensive and time-consuming. Generally, the electrodeposition method represents a lower cost technique being well adapted towards industrial applications so that it is considered suitable to be applied to build large-area super-hydrophobic surfaces [14].

Until now, in spite of the various preparation techniques, the commercialization of super-hydrophobic surfaces have been significantly limited due to their poor mechanical abrasion resistance and surface chemical stability, as well as the fragility of their microscopic roughness features [15]. Mechanical strength and durability of super-hydrophobic coatings are, therefore, of considerable focus in recent studies [15,16]. The robust super-hydrophobic coating of contact angle above 163° was fabricated by She et al. on a magnesium alloy substrate via the combination of electrodeposition and chemical modification technique [17]. After mechanical abrasion for 0.7 m with the 800 grit SiC sandpapers under applied pressure of 1.2 kPa, the as-prepared sample with pinecone-like hierarchical structure could maintain the contact angle above 150° [17]. Electrodeposition of Mg–Mn–Ce magnesium plate in the ethanol solution of cerium nitrate and myristic acid was used by Liu et al. to construct a super-hydrophobic surface with a maximum contact angle of 158° that lose the super-hydrophobicity after abrasion for 0.4 m under the pressure of 1.3 kPa [18]. A super-hydrophobic Ni coating with pinecone-like hierarchical micro-nanostructure was prepared by Su et al. [19], which involved electroplating in a Watts bath and heat-treatment with triethoxysilane (AC-FAS). The deposit was found to possess super-hydrophobicity and a good mechanical abrasion resistance after mechanical abrasion against 800 grit SiC sandpaper for 1.0 m under the pressure of 4.80 kPa [19]. Jain et al. [20] prepared a durable copper-based super-hydrophobic surface with cauliflower shaped fractal morphology via an electrodeposition route which lost its non-wetting nature after being dragged on 800 grit sandpaper for a distance of 2.0 m under the applied pressure of 3 kPa. Tam et al. fabricated super-hydrophobic nanocrystalline Ni–PTFE composite coating by co-electrodeposition process [21]. On the 800 grit sandpaper, the water contact angle of the Ni–PTFE composite coating can remain above 150° after 50 m of abrasion under the applied pressure of 2.0 kPa [21]. Until now, the abrasion resistance of super-hydrophobic surfaces is still not satisfactory, thus more research attention should be paid to improve its mechanical durability.

In this work, robust super-hydrophobic Co–Ni coating with cauliflower-shaped micro-nano structures was fabricated on carbon steel via a low-cost electrochemical deposition process. In the linear abrasion test, the super-hydrophobic properties can be maintained after abrasion distance of 12 m under the applied pressure of 5 kPa, which was significantly improved compared with the previous reports [19,20]. Moreover, electrochemical tests demonstrated that the super-hydrophobic Co–Ni coating possessed good corrosion resistance for carbon steel substrate.

2. Materials and Methods

Nickel chloride ($NiCl_2$), cobalt chloride ($CoCl_2$) and boric acid (H_3BO_3) were bought from Sino pharm Chemical Reagent Co., Ltd. (Beijing, China). The low surface energy material (1H,1H,2H,2H-Perfluorooctyltrichlorosilane, PFTEOS) was provided from Beijing Bai Ling Wei Technology Co., Ltd. (Beijing, China). The chemical reagents were of analytical grade and used as-received.

A conventional three electrode configuration was used to prepare the samples. Carbon steel substrate in a square shape was used as working electrode. A platinum sheet with a size of 30×30 mm^2 worked as counter electrode while saturated calomel electrode (SCE) was used as the reference electrode. Before electrochemical deposition, the carbon steel substrate was sealed in epoxy with an exposed surface area of 1 cm^2. Afterwards, it was mechanically polished down to 2000 grit size by SiC sandpapers and rinsed with deionized water. The mixed solution was selected based on our previous report [22] and improved for the purpose of abrasion resistance. The solutions with $CoCl_2$ (0.1 mol/L), $NiCl_2$ (0.03 mol/L) and H_3BO_3 (0.1 mol/L) were prepared and used directly. The cyclic voltammograms of carbon steel were recorded in the solution of $CoCl_2$ 0.1 mol/L + H_3BO_3 0.1 mol/L, $NiCl_2$ 0.03 mol/L + H_3BO_3 0.1 mol/L, $CoCl_2$ 0.1 mol/L + $NiCl_2$ 0.03 mol/L + H_3BO_3 0.1 mol/L with a scan rate of 10 mV/s. The Co–Ni coatings were deposited at a constant potential of −1.0, −1.4 and −1.7 V for 3000 s at room temperature in mixed solution $CoCl_2$ 0.1 mol/L + $NiCl_2$ 0.03 mol/L + H_3BO_3 0.1 mol/L. After electrochemical deposition, the as-prepared samples were treated with 5 wt.% PFTEOS ethanol solution for 1 h at room temperature, then rinsed and dried for investigations.

To test the wetting property of the as-prepared samples, the water contact angle and water sliding angle (WSA) were measured by an Automatic Contact Angle Meter (Model SL150 Series, USA

KINO, Boston, MA, USA) at ambient temperature using 5 and 12 µL distilled water, respectively. During the measurement of water sliding angle, the inclination of the slope was controlled by the cornering device. For accuracy, the different samples were measured under the same condition for three times in order to obtain the average values reported. The surface morphologies of all samples were obtained by scanning electron microscope (SEM, EVO MA 25/LS 25, Carl Zeiss, Opokochen, Germany). The chemical compositions and chemical states of the top coating surface were characterized by EDS (X-flash-Detector 5010, Bruker, Karlsruhe, Germany) and X-ray photoelectron spectrometer (XPS, Kratos Axis Ultra DLD, Kratos Analytical, Hadano, Japan). The linear abrasion test with 800 grit SiC sandpaper was carried out to evaluate the mechanical durability of the as-prepared samples [23]. The laser scanning confocal microscopy (LSCM, LEXTOLS4000 OLYMPUS, Tokyo, Japan) was employed to investigate the surface roughness of the as-deposited coatings.

The electrochemical corrosion resistance was evaluated by a conventional three electrode system, in which the as-prepared sample, a platinum plate and a saturated calomel electrode (SCE) were used as the working electrode, counter electrode and reference electrode, respectively. The test was conducted with an electrochemical workstation (Gamry-Reference 3000, Gamry Instruments, Warminster, PA, USA) in 3.5 wt.% NaCl solution at room temperature. The potentiodynamic polarization curves were recorded from -1.3 to 0.0 V with a scan rate of 0.5 mV/s. The Tafel extrapolation method was used to extract the corrosion potential (E_{corr}) and corrosion current density (I_{corr}) from the polarization curves. Electrochemical impedance spectroscopy (EIS) measurements were performed at open circuit potential (OCP) under the employed amplitude signal of 5 mV in the frequency range from 10^5 to 10^{-2} Hz. Software ZSimpWin (version 3.3) was utilized to fit the obtained impedance data.

3. Results and Discussion

3.1. Surface Morphology of the As-Deposited Co–Ni Coating

In order to determine the applied deposition potentials, the cyclic voltammograms of carbon steel were recorded with a scan rate of 10 mV/s in different solutions (Figure S1). Accordingly, the applied potentials for metal electrodeposition should be more negative than its equilibrium potential. The higher over-potential will lead to larger deposition driving force, which will favor the formation of hierarchical structures. Figure 1 shows typical morphologies of the bare carbon steel and electrodeposited Co–Ni coating under different applied potentials for 3000 s in the mixed solution with CoCl$_2$ (0.1 mol/L), NiCl$_2$(0.03 mol/L) and H$_3$BO$_3$ (0.1 mol/L). For the bare carbon steel, the polished surface displayed lots of scratches (Figure 1a). After electrochemical deposition under the applied potential at -1.0 V for 3000 s in the above-mentioned solution, the carbon steel surface was covered by uniform granular structures containing average size of sub-micrometer in diameter and the as-polished scratches were covered completely (Figure 1b). As shown in the cross-section view, the thickness of the coating is around 26 µm for the deposition time of 3000 s, and the EDS result showed that the cobalt content is around 93.8%, which is higher than that of the initial electrolyte (77%, CoCl$_2$ 0.1 mol/L NiCl$_2$ 0.03 mol/L). After the deposition at -1.4 V for 3000 s, the coating surface evidenced lots of spherical humps with an average size of 33 µm and a large number of cracks (Figure 1c), which may be attributed to the internal stress generated during the electrochemical deposition process. The cross-section image revealed that the thickness of the coating was around 40 µm and the large spherical humps grew from the thin layer composed of small irregular crystals. Increasing the applied potential to -1.7 V, the cauliflower-shaped micro-nano structures with multiscale fractal nature were obtained (Figure 1d). The EDS result shows that the cobalt content decreased to around 77.7% as presented in Table 1, which is similar to that of the initial electrolyte (77%). The anomalous Co–Ni deposition behavior can be attributed to the formation of Co hydroxyl precipitate, which could hinder the subsequent Ni deposition at the solid/electrolyte interface [24]. Under different thermodynamic and kinetic conditions, the various external crystal morphologies including powder [25], film [26] and dendrite shapes [27,28] could be generated during the electrodeposition process. In our case, increasing the over-potential can favor the development of protrusions in the direction of increasing concentration, therefore leading to the formation of cauliflower-shaped micro-nano structures.

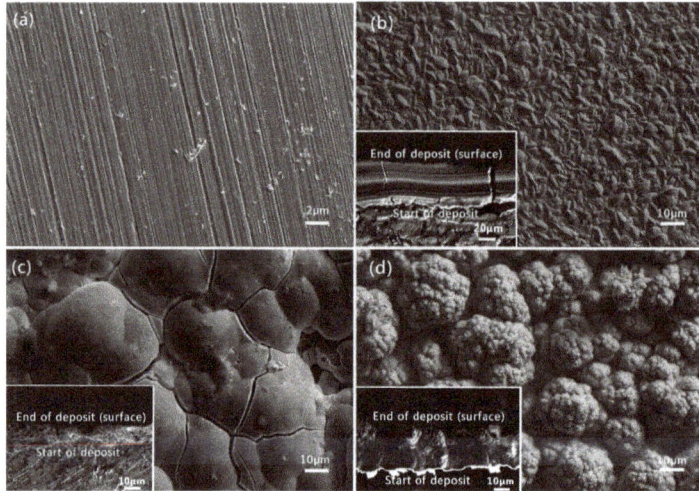

Figure 1. SEM images of (**a**) the bare carbon steel and the Co–Ni coatings under the applied potential of (**b**) −1.0 V, (**c**) −1.4 V, and (**d**) −1.7 V for 3000 s in the mixed solution at room temperature. The insets show the corresponding cross section.

Table 1. EDS results of the deposited Co–Ni coatings under different applied potentials for 3000 s in the mixed solution.

Potentials	Coating	
	Ni (at.%)	Co (at.%)
−1.0 V	6.2	93.8
−1.4 V	19.0	81.0
−1.7 V	22.3	77.7

3.2. Structure of As-Deposited Co–Ni Coating

To identify the crystal structure of the deposited coatings, the XRD technique was conducted. Figure 2 shows the XRD patterns of the bare carbon steel substrate and Co–Ni coatings prepared at the applied potentials of −1.0, −1.4 and −1.7 V, respectively. For the bare carbon steel substrate, the diffraction peaks in 2θ of 44.5°, 65° and 82° can be attributed to the existence of Fe (JCPDS file No. 870721) (Figure 2a). After electrochemical deposition, the diffraction peaks located at 44.5°, 75.9°, and 84.1° correspond to the peaks of (002), (110), and (103) crystalline faces of close-packed hexagonal (hcp) Co (JCPDS file No. 040850) (Figure 2b–d); the diffraction peaks near 44.2°, 51.5°, and 75.9° can be assigned to the peaks of (111), (200), and (220) crystalline faces of face-centered cube (fcc) Co (JCPDS file No. 897093) (Figure 2b–d). In the XRD patterns, no pure nickel can be detected, revealing the formation of homogeneous solid solutions of Co–Ni alloy under the deposition conditions [29,30]. When deposited at the applied potential of −1.0 V, the Co–Ni coating shows completely a hexagonal structure with the cobalt content of around 93.8% according to the EDS results (Table 1). By increasing the applied deposition potential, the cobalt content decreases. As the cobalt content on the coating obtained at the applied potential of −1.7 V decreases to around 77.7%, the face-centered cubic structure becomes dominant.

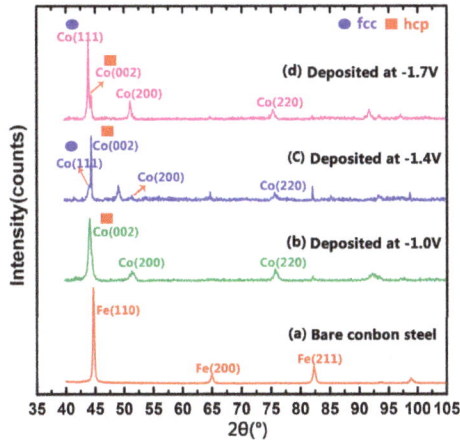

Figure 2. XRD patterns of (**a**) the bare carbon steel surface and the Co–Ni coatings for 3000 s under the applied potential of (**b**) −1.0 V, (**c**) −1.4 V, and (**d**) −1.7 V in the mixed solution at room temperature.

3.3. Surface Wetting Property of As-Deposited Co–Ni Coatings

Before the wetting property measurement of the as-deposited Co–Ni coatings, the surface roughness of these coatings was evaluated by the laser scanning confocal microscopy. As shown in the Figure S2, the Co–Ni coating deposited at −1.4 V showed the R_a of around 1.71 μm, which was higher than that of Co–Ni coating deposited at −1.0 V. Meanwhile, the Co–Ni coating with cauliflower-shaped micro-nano structures deposited at −1.7 V showed the highest surface roughness with R_a of 7.77 μm. By adjusting the applied potential to a more negative direction, hierarchical structures with higher surface roughness can be favored to be generated.

To examine the surface wetting property of the as-prepared Co–Ni coatings, the water contact angle tests were conducted. Figure 3 depicts the contact angle variations of the Co–Ni coatings with the applied potentials before and after modification by PFTEOS in ethanol solution. For bare carbon steel substrate after polishing, and Co–Ni coating deposited at the applied potential of −1.0 V, the contact angles were around 20°. After modification by PFTEOS, the water contact angle reached 95° and the surfaces displayed a hydrophobic property. After electrodeposition at the applied potentials ranging from −1.4 to −1.7 V, the water contact angles of the Co–Ni coatings with spherical humps structures and cauliflower-shaped micro-nano structures decreased to almost zero, showing the super-wetting properties of deposited Co–Ni coatings. However, after modification by 5 wt.% PFTEOS in ethanol solution for 1 h at room temperature, the water contact angle increases drastically to 140° for the sample with spherical humps structures, and to 161° for the sample with cauliflower-shaped micro-nano structures deposited at −1.7 V, respectively.

The above results indicate that the wetting properties of the deposited hierarchical Co–Ni coating with cauliflower-shaped micro-nano structures converted from super-wetting to super-hydrophobic behaviors during the PFTEOS modification process. The super-hydrophobic behavior on Co–Ni coating with cauliflower-shaped micro-nano structures was endowed with a high surface roughness and low surface energy materials achieved by the combination of electrodeposition process at higher overpotential and surface modification by PFTEOS. The super-hydrophobicity of the as-deposited Co–Ni coating can be explained by the Cassie-Baxter model. Based on the Cassie-Baxter Equation [31]:

$$\cos \theta = f_{sl}\left(\cos \theta_y + 1\right) - 1 \tag{1}$$

where, θ_y and θ are Young's contact angle and liquid–gas contact angle, f_{sl} is the contact area fraction of solid–liquid interface, the area fraction of water-air interface calculated is around 94%, suggesting that the water droplet was sustained by the heterogeneous composite surface consisting of Co–Ni cauliflower-shaped micro-nano structures and air cushion among these structures.

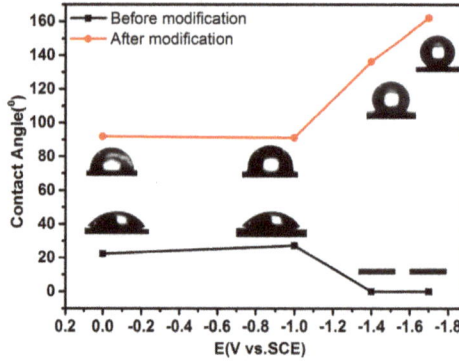

Figure 3. The water contact angles of the samples deposited at different potentials before and after modification by PFTEOS.

3.4. Surface Composition of the Super-Hydrophobic Co–Ni Coating

X-ray photoelectron spectroscopy (XPS) was utilized to confirm the adsorption of the PFTEOS molecules on the Co–Ni coating with cauliflower-shaped micro-nano structures deposited at the applied potential of –1.7 V. As can be seen in Figure 4, before PETEOS treatment, the strong signals of Co $2p$ and Ni $2p$ core levels reveals that the formation of Co–Ni electrodeposits (Figure 4a–c). After treatment by PFTEOS ethanol solution for 1 h, strong signal of F $1s$ and Si $2p$ core levels can be observed (Figure 4a,d,e). Figure 4f depicted the high-resolution C $1s$ core level spectra. The peaks at binding energy of 293.6, 291.1 and 288.7 eV can be ascribed to the –C–F$_3$ group, –C–F$_2$– group and –C–CF$_2$– group respectively [19,32]. These results demonstrated that the PFTEOS molecules adsorbed successfully on the Co–Ni coating with cauliflower-shaped micro-nano structures after treatment. Accordingly, the Co–Ni coating with cauliflower-shaped micro-nano structures prepared by electrochemical deposition exhibits higher surface roughness R_a of 7.77 μm. After PFTEOS treatment, the fluorinated components with low surface energy endowed the rough Co–Ni coating super-hydrophobic properties, which is similar to the observed lotus leaves with dual micro and nano-scale structure covered by a wax layer [3].

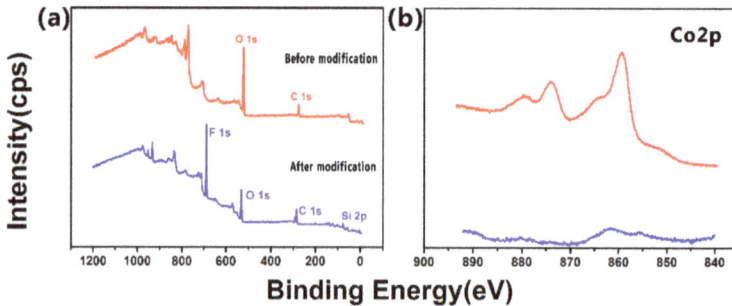

Figure 4. *Cont.*

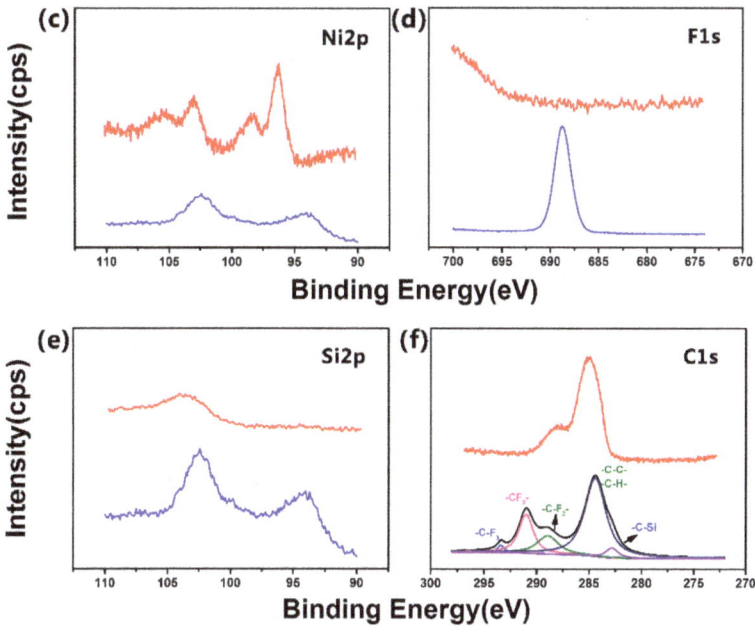

Figure 4. XPS spectra before and after modification of the Co–Ni coating deposited under the applied potential at −1.7 V in the mixed solution: (**a**) XPS survey spectra; (**b**) Co 2*p* spectra; (**c**) Ni 2*p* spectra; (**d**) F 1s spectra; (**e**) Si 2*p* spectra; and (**f**) high resolution C 1s spectra.

3.5. Abrasion Resistance of the Super-Hydrophobic Co–Ni Coating

Due to the special features (such as dual micro-nano structures and modification by low surface energy materials) which is essential for fabricating super-hydrophobic surfaces, such surfaces are susceptible to mechanical abrasion. At present, to enhance the abrasion resistance of super-hydrophobic coatings has become the main concern for their practical applications [23]. Recently, Ras et al. [23] suggested that linear abrasion should be adopted to assess the mechanical durability of the super-hydrophobic coating because this wear-test method is accessible to most lab researches, applicable in most industrial production, and able to generate a large uniformly surface suitable for wetting measurements. Therefore, in this work the linear abrasion test was conducted, which is shown in Figure 5. The 800 grit SiC sandpaper was placed face up and used as abrasion surface. The super-hydrophobic Co–Ni coating was tested under the applied pressure of 5 kPa at a speed of 5 mm/s.

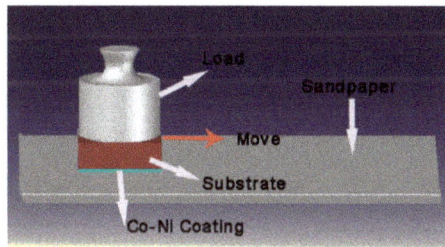

Figure 5. A scheme for linear abrasion test.

Figure 6 depicts the SEM images for the super-hydrophobic Co–Ni coating deposited at −1.7 V before and after linear abrasion tests and the relationship between contact angles and abrasion distance. As shown in Figure 6a, the Co–Ni coating with cauliflower-shaped micro-nano structures shows contact angle of 161° and water sliding angle of 1° before linear abrasion tests, exhibiting greater super-hydrophobicity. After abrasion distance of 1.5 m under the applied pressure of 5 kPa, few prominent cauliflower-shaped micro-nano structures were worn and obvious scratches appeared on the top of these structures (Figure 6b). It is obvious that the cauliflower-shaped micro-nano structures were well preserved and the water contact angle maintained around 160° with water sliding angle of 2°. After increasing the abrasion distance to 6 m, large cauliflower-shaped structures were worn more seriously, and the water contact angle dropped to 158° and the water sliding angle increased to 5° (Figure 6c). When abrasion distance reaching 12 m, all the large cauliflower-shaped microstructures appeared different degrees of wear and the water contact angle dropped dramatically to near 150° (Figure 6d). With the increase of abrasion distance, the water sliding angle increases to 8° after 12 m of abrasion. When the abrasion distance increased to 24 m, the large cauliflower-shaped microstructures were almost completely worn away. The water contact angle dropped to near 143° and the sliding angle increased to 12° (Figure 6f). With the increase of abrasion distance, the wetting properties of deposited hierarchical Co–Ni coating with cauliflower-shaped micro-nano structures transited from super-hydrophobic behaviors to hydrophobic behaviors.

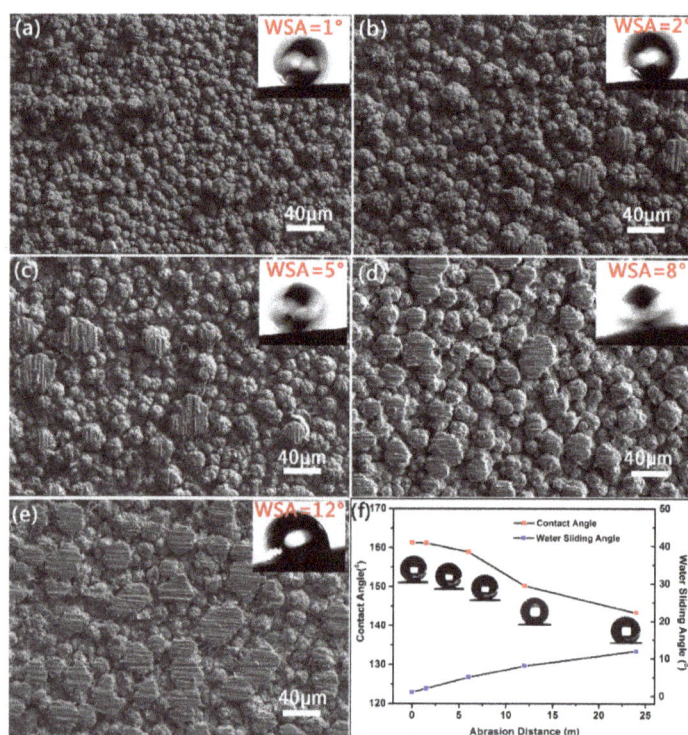

Figure 6. SEM images for (**a**) super-hydrophobic Co–Ni coating before abrasion and after abrasion under the applied pressure of 5 kPa for (**b**) 1.5 m, (**c**) 6 m, (**d**) 12 m, and (**e**) 24 m; (**f**) The water contact angle and water sliding angle variations on these surface with the abrasion distance. The insert images in (**a**) to (**e**) are the profiles of a water droplet sliding on the Co–Ni coating with different water sliding angles.

Moreover, the mechanical durability, which was tested under the similar conditions by the linear abrasion method, of the electrodeposited super-hydrophobic coatings including nickel [19], copper [20] and cobalt [33] was summarized in Table 2. As compared to the electrodeposited coatings in the literatures, our Co–Ni coating with cauliflower-shaped micro-nano structures exhibits longer abrasion distance before the water contact angle drops to 150° except the Ni–PTFE composite coating tested under lower pressure. When the as-prepared Co–Ni coating loses its super-hydrophobic properties, the water sliding angle increases to 8°, which is better than other electrodeposited coatings. These results demonstrated that our prepared super-hydrophobic Co–Ni coating has great abrasion resistance. And the great abrasion resistance can be attributed to the high cobalt content which possesses higher hardness. This comparison implies that, to further increase the abrasion resistance of super-hydrophobic coating, the co-electrodeposition of Co–Ni coating with PTFE particles maybe a good choice.

Table 2. Abrasion distances leading to the loss of super-hydrophobic properties. Abrasive medium: 800 grit SiC sandpaper.

Materials	Pressure (kPa)	Abrasion Length (m)	Initial WCA (°)	Final WCA (°)	Initial WSA (°)	Final WSA (°)
Co–Ni coating (this work)	5.0	12.0	161	150	1	8
Ni–PTFE [21]	2.0	50.0	156	150	3	52
Cu [20]	3.0	2.0	162	143	3	18
Ni [19]	4.8	1.0	162	150	3	15
Mg–Mn–Ce [18]	1.3	0.4	160	150	2	Not given
Co [33]	1.5	1.1	156	148	1	40
Cu compound [34]	1.2	0.7	163	140	1	Not given

3.6. Corrosion Resistance of the Super-Hydrophobic Co–Ni Coating

The anti-corrosion performance of the super-hydrophobic Co–Ni coating was evaluated by electrochemical methods. Figure 7 depicts the potentiodynamic polarization curves of carbon steel substrate, Co–Ni coatings deposited under the applied potentials of −1.0 and −1.4 V, and super-hydrophobic Co–Ni coating obtained at the applied potentials of −1.7 V with a sweep rate of 0.5 mV/s in 3.5 wt.% NaCl solution after immersion for 1 h, respectively. Tafel extrapolation method was applied to extract the corrosion potential (E_{corr}) and corrosion current density (I_{corr}), which was summarized in Table 3.

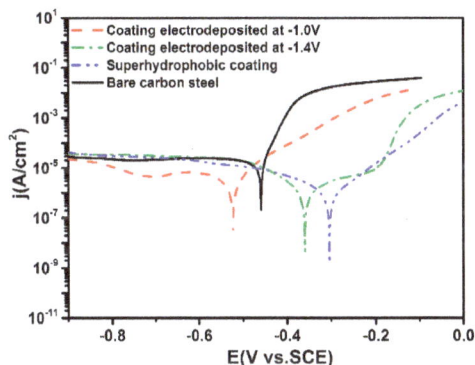

Figure 7. Potentiodynamic polarization curves of the bare carbon steel, Co–Ni coating deposited at −1.0 and −1.4 V, and super-hydrophobic Co–Ni coating deposited at −1.7 V after immersion in 3.5 wt.% NaCl aqueous solution for 1h with a scan rate of 0.5 mV/s.

Table 3. Extracted corrosion potential (E_{corr}) and current density (I_{corr}) from the potentiodynamic polarization measurements presented in Figure 6.

Sample	E_{corr} (mV)	I_{corr} (A/cm^2)	Corrosion Rate (mm/a)
Bare carbon steel	−459.4	1.23×10^{-5}	0.137
Co–Ni coating at −1.0 V	−522.7	5.08×10^{-6}	0.057
Co–Ni coating at −1.4 V	−359.4	1.50×10^{-6}	0.017
Super–hydrophobic coating	−303.3	5.87×10^{-7}	0.0066

Revealed from the Figure 7, the carbon steel substrate displayed negative corrosion potential (E_{corr} = −459 mV vs. SCE) with high corrosion current density (I_{corr} = 1.23 × 10^{-5} A/cm^2). After electrodeposition under the applied potential of −1.0 V, the corrosion potential of Co–Ni coating shifted negatively to −522 mV, which indicated that the Co–Ni coating with uniform granular structures displayed a higher corrosion susceptibility compared with the carbon steel substrate. The corrosion potential of Co–Ni coating deposited at −1.4 V shifted 100 mV toward positive direction compared with carbon steel with lower corrosion current density of 1.50 × 10^{-6} A/cm^2. For the super-hydrophobic Co–Ni coating deposited at −1.7 V, the corrosion potential and corrosion current density are −303 mV and 5.87 × 10^{-7} A/cm^2 respectively. The corrosion current density of as-prepared super-hydrophobic Co–Ni coating was 20 times lower than that of bare carbon steel substrate, demonstrating the significantly improved anti-corrosion performance of the super-hydrophobic Co–Ni coating with cauliflower-shaped micro-nano structures.

To further evaluate the corrosion resistance of the superhydrophobic Co–Ni coating, electrochemical impedance spectroscopy (EIS) was applied. Figure 8 displays the EIS results of the carbon steel substrate, Co–Ni coatings deposited at −1.0 and −1.4 V, and super–hydrophobic Co–Ni coating at −1.7 V in the form of Nyquist plots and Bode plots. From Figure 8a,b, it can be noted that the super–hydrophobic Co–Ni coating exhibits a larger semicircle in Nyquist plots, revealing a higher charge transfer resistance compared with other Co–Ni coatings. Figure 8c shows the corresponding impedance modulus versus frequency plot. The impedance modulus value of the super–hydrophobic Co–Ni coating (4.6 × 10^5 Ω cm^2) is larger than those of the Co–Ni coating deposited at −1.4 V (4.5 × 10^4 Ω cm^2), Co–Ni coating deposited at −1.0 V (4.2 × 10^3 Ω cm^2) and carbon steel substrate (3.5 × 10^3 Ω cm^2) at the frequency of 0.01 Hz. The higher impedance modulus at low frequency domain could be attributed to the super–hydrophobic behavior of the Co–Ni coating, which barriers the infiltration of aggressive medium liquid into the substrate, indicating a better anti–corrosion performance of the as–prepared super–hydrophobic Co–Ni coating. The impedance modulus of the coatings decreased with the increase of the frequency, finally reached to a value of around 60 Ω cm^2 at the high frequency domain which indicated that the coatings could allow the aggressive species inside the electrolyte infiltrate the coatings easily. As it can be seen from Figure 8d, two time constants exist within the testing frequency range for the super–hydrophobic Co–Ni coating, corresponding to the electrochemical process at high frequency range and the film/metal interface at low frequency. For the other Co–Ni coatings and carbon steel substrate, only one time constant can be observed, which reveals only one electrochemical process occurring within the testing frequency range.

In order to further understand the electrochemical process, equivalent circuit models were applied to fit EIS results. The simple Randles circuit shown in Figure 8e was used to analyze the EIS results of carbon steel substrate, where R_s and R_{ct} stand for the solution resistance and the charge transfer resistance respectively, the constant phase element CPE$_{dl}$ models the non-ideal capacitance at solid/electrolyte interface as a result of the inhomogeneous current distribution. According to the literature, the impedance of CPE was defined as follows:

$$Z_{CPE} = 1/Y_0(j\omega)^n \tag{2}$$

where Z_{CPE} represents CPE impedance, Y_0 represents modulus, ω represents angular frequency and n is the exponent of the CPE varying between 0 and 1. The equivalent electrical circuit shown in Figure 8f was used to analyze the EIS results of the Co–Ni coatings deposited at −1.0 and −1.4 V as well as the super-hydrophobic Co–Ni coating deposited at −1.7 V, where R_c and CPE_c represent the coating resistance and non-ideal capacitance at coating/electrolyte interface, respectively. The electrochemical parameters acquired through simulating with the equivalent circuit are presented in Table 4. From the obtained results, the R_{ct} value of the super-hydrophobic Co–Ni coating is three orders of magnitude higher than that of carbon steel substrate. Generally, the inhibition efficiency (η) was computed with the equation as follows [35]: $\eta(\%) = (1 − R_{ctb}/R_{ctc}) \times 100\%$, where the R_{ctb} and R_{ctc} are the charge transfer resistance of bare carbon steel and the super-hydrophobic Co–Ni coating, respectively. Based on the values in Table 4, the inhibition efficiency of the as-deposited super-hydrophobic Co–Ni coating can be calculated as 99.3%, suggesting that the electrochemical charge transfer process of super-hydrophobic Co–Ni coating is very difficult to occur under the tested conditions.

The EIS results were in agreement with the potentiodynamic polarization curves, which indicates that the as-prepared super-hydrophobic Co–Ni coating with cauliflower-shaped micro-nano structures has a larger charge transfer resistance and exhibits excellent anti-corrosion performance. The lower corrosion probability and corrosion rate of the super-hydrophobic Co–Ni coating could be attributed to the trapped air among the hierarchical cauliflower-shaped micro-nano structures, which can work as corrosion barrier effectively through limiting the lower contact area of solid coating with the electrolyte solution at the interface, therefore providing better protection for the carbon steel substrate.

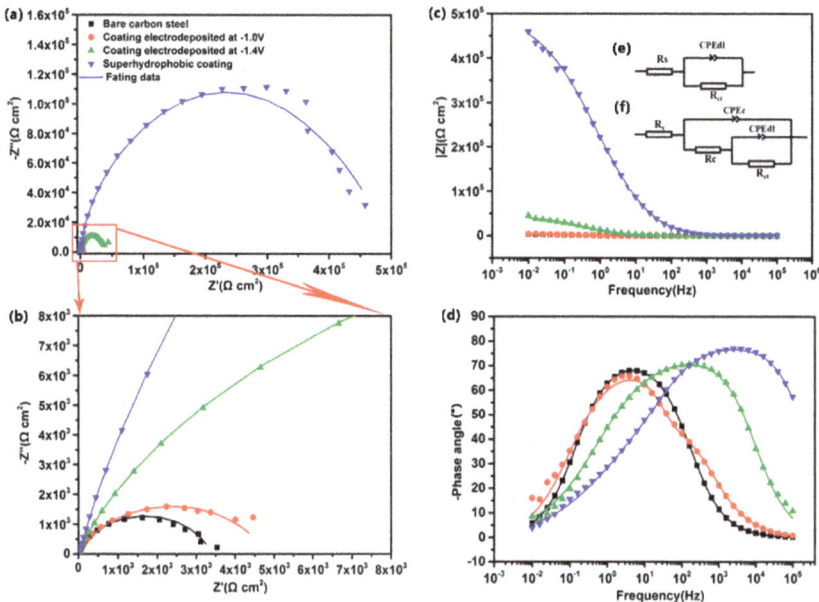

Figure 8. EIS results of the bare carbon steel, Co–Ni coating deposited at −1.0 and −1.4 V, and super-hydrophobic Co–Ni coating deposited at −1.7 V in 3.5 wt.% NaCl solution after immersion for 1 h: (**a**) Nyquist plots; (**b**) high-resolution Nyquist plot in (**a**) marked with red box; (**c**) bode −|Z| versus frequency plots; and (**d**) bode-phase angle versus frequency plots. (**e**) Equivalent circuits for simulating EIS results of bare carbon steel. (**f**) Equivalent circuits for simulating EIS results of the Co–Ni coating deposited at −1.0 and −1.4 V, and the super-hydrophobic Co–Ni coating deposited at −1.7 V.

Table 4. The impedance parameters of bare carbon steel, Co–Ni coating deposited at −1.0 and−1.4 V, and super-hydrophobic Co–Ni coating deposited at −1.7 V extracted by fitting the EIS results recorded in 3.5 wt.% NaCl solution (Figure 8) with the corresponding equivalent circuits. The units of R_s, R_{ct}, R_c and CPE are $\Omega \cdot cm^2$, $k\Omega \cdot cm^2$, $k\Omega \cdot cm^2$ and $\Omega^{-1} \cdot s^n \cdot cm^{-2}$, respectively.

Fitted Parameters	R_s	CPE_{dl}	n_{dl}	R_{ct}	CPE_c	n_c	R_c
Bare carbon steel	8.42	1.08×10^{-3}	0.659	2.40	–	–	–
Co–Ni coating deposited at −1.0 V	10.75	3.59×10^{-5}	0.983	4.65	2.83×10^{-4}	0.729	0.116
Co–Ni coating deposited at −1.4 V	11.68	3.06×10^{-5}	0.629	26.4	1.00×10^{-5}	0.816	16.1
Super–hydrophobic coating	23.82	1.43×10^{-6}	0.582	369.1	1.78×10^{-7}	0.868	98.9

4. Conclusions

Mechanical durability and corrosion resistance are of great importance for the real industrial applications of super-hydrophobic surfaces. In our work, a super-hydrophobic coating with a contact angle around 161° was successfully fabricated on a carbon steel substrate by electrochemical deposition. The super-hydrophobicity of the constructed coating derived from the cauliflower-shaped micro-nanostructures and the low-surface-energy PFTEOS layer with fluorinated components. In addition, the Co–Ni coatings possessing different surface microstructures were prepared using electrochemical deposition by varying the applied potentials at room temperature. The as-prepared super-hydrophobic Co–Ni coating with cauliflower-shaped micro-nanostructures exhibited excellent mechanical abrasion resistance. The linear abrasion test revealed that the super-hydrophobic properties were maintained well after abrasion distance of 12 m under the applied pressure of 5 kPa, which was significantly improved by the increase of cobalt content in the Co–Ni coating. Furthermore, electrochemical tests indicated that the super-hydrophobic Co–Ni coating showed lower corrosion current density and a higher charge transfer resistance as a result of limiting the contact area of the solid coating with the electrolyte solution at the interface, providing excellent protection for the carbon steel. In summary, through a combination of electrodeposition and the modification by low surface energy materials, this facile and low-cost preparation process offers an effective technique and promises the practical applications of super-hydrophobic coatings.

Supplementary Materials: The following are available online at http://www.mdpi.com/2079-6412/9/4/232/s1, Figure S1: Cyclic voltammograms of carbon steel obtained with a scan rate of 10 mV/s in the solution of $CoCl_2$ 0.1 mol/L + H_3BO_3 0.1 mol/L, $NiCl_2$ 0.03 mol/L + H_3BO_3 0.1 mol/L, $CoCl_2$ 0.1 mol/L + $NiCl_2$ 0.03 mol/L + H_3BO_3 0.1 mol/L; Figure S2: Surface roughness of Co–Ni coatings deposited under the applied potentials of (a) −1.0 V, (b) −1.4 V, and (c) −1.7 V for 3000 s in the mixed solution at room temperature.

Author Contributions: Conceptualization, Y.X.; Methodology, Y.X.; Software, S.W.; Validation, S.W. and P.B.; Formal Analysis, S.W.; Investigation, S.W. and P.B.; Resources, S.W.; Data Curation, G.Z.; Writing—Original Draft Preparation, S.W.; Writing—Review and Editing, Y.X. and Y.J.; Visualization, Y.X.; Supervision, Y.J.; Project Administration, Y.X.; Funding Acquisition, Y.X.

Funding: This research was funded by the Fundamental Research Funds for the Central Universities China (Project ID: FRF-TP-18-009A1).

Conflicts of Interest: The authors declare no conflict of interest.

References

1. Sanchez, C.; Arribart, H.; Guille, M.M.G. Biomimetism and bioinspiration as tools for the design of innovative materials and systems. *Nat. Mater.* **2005**, *4*, 277–288. [CrossRef] [PubMed]
2. Serrà, A.; Zhang, Y.; Sepúlveda, B.; Gómez, E.; Nogués, J.; Michler, J.; Philippe, L. Highly active ZnO-based biomimetic fern-like microleaves for photocatalytic water decontamination using sunlight. *Appl. Catal. B Environ.* **2019**, *248*, 129–146. [CrossRef]
3. Liu, K.; Jiang, L. Bio-inspired self-cleaning surfaces. *Annu. Rev. Mater. Res.* **2012**, *42*, 231–263. [CrossRef]
4. Liu, K.; Yao, X.; Jiang, L. Recent developments in bio-inspired special wettability. *Chem. Soc. Rev.* **2010**, *39*, 3240. [CrossRef] [PubMed]

5. Vazirinasab, E.; Jafari, R.; Momen, G. Application of superhydrophobic coatings as a corrosion barrier: A review. *Surf. Coat. Technol.* **2018**, *341*, 40–56. [CrossRef]

6. Eberle, P.; Tiwari, M.K.; Maitra, T.; Poulikakos, D. Rational nanostructuring of surfaces for extraordinary icephobicity. *Nanoscale* **2014**, *6*, 4874–4881. [CrossRef] [PubMed]

7. Ma, Q.; Cheng, H.; Fane, A.G.; Wang, R.; Zhang, H. Recent development of advanced materials with special wettability for selective oil/water separation. *Small* **2016**, *12*, 2186–2202. [CrossRef]

8. Pu, X.; Li, G.; Huang, H. Preparation, anti-biofouling and drag-reduction properties of a biomimetic shark skin surface. *Biol. Open* **2016**, *5*, 389–396. [CrossRef]

9. Liao, R.; Zuo, Z.; Guo, C.; Yuan, Y.; Zhuang, A. Fabrication of superhydrophobic surface on aluminum by continuous chemical etching and its anti-icing property. *Appl. Surf. Sci.* **2014**, *317*, 701–709. [CrossRef]

10. Long, J.; Fan, P.; Gong, D.; Jiang, D.; Zhang, H.; Li, L.; Zhong, M. Superhydrophobic surfaces fabricated by femtosecond laser with tunable water adhesion: From lotus leaf to rose petal. *ACS Appl. Mater. Interfaces* **2015**, *7*, 9858–9865. [CrossRef]

11. Yu, J.; Qin, L.; Hao, Y.; Kuang, S.; Bai, X.; Chong, Y.; Zhang, W.; Wang, E. Vertically aligned boron nitride nanosheets: Chemical vapor synthesis, ultraviolet light emission, and superhydrophobicity. *ACS Nano* **2010**, *4*, 414–422. [CrossRef]

12. Jeong, C.; Choi, C.H. Single-step direct fabrication of pillar-on-pore hybrid nanostructures in anodizing aluminum for superior superhydrophobic efficiency. *ACS Appl. Mater. Interfaces* **2012**, *4*, 842–848. [CrossRef] [PubMed]

13. Lee, M.; Kwak, G.; Yong, K. Wettability control of ZnO nanoparticles for universal applications. *ACS Appl. Mater. Interfaces* **2011**, *3*, 3350–3356. [CrossRef]

14. Tam, J.; Palumbo, G.; Erb, U. Recent advances in superhydrophobic electrodeposits. *Materials* **2016**, *9*, 151. [CrossRef]

15. Verho, T.; Bower, C.; Andrew, P.; Franssila, S.; Ikkala, O.; Ras, R.H.A. Mechanically durable superhydrophobic surfaces. *Adv. Mater.* **2011**, *23*, 673–678. [CrossRef]

16. Peng, C.; Chen, Z.; Tiwari, M.K. All-organic superhydrophobic coatings with mechanochemical robustness and liquid impalement resistance. *Nat. Mater.* **2018**, *17*, 355–360. [CrossRef] [PubMed]

17. She, Z.; Li, Q.; Wang, Z.; Li, L.; Chen, F.; Zhou, J. Researching the fabrication of anticorrosion superhydrophobic surface on magnesium alloy and its mechanical stability and durability. *Chem. Eng. J.* **2013**, *228*, 415–424. [CrossRef]

18. Liu, Q.; Chen, D.; Kang, Z. One-step electrodeposition process to fabricate corrosion-resistant superhydrophobic surface on magnesium alloy. *ACS Appl. Mater. Interfaces* **2015**, *7*, 1859–1867. [CrossRef] [PubMed]

19. Su, F.; Yao, K. Facile fabrication of superhydrophobic surface with excellent mechanical abrasion and corrosion resistance on copper substrate by a novel method. *ACS Appl. Mater. Interfaces* **2014**, *6*, 8762–8770. [CrossRef]

20. Jain, R.; Pitchumani, R. Facile fabrication of durable copper-based superhydrophobic surfaces via electrodeposition. *Langmuir* **2018**, *34*, 3159–3169. [CrossRef]

21. Tam, J.; Jiao, Z.; Lau, J.C.F.; Erb, U. Wear stability of superhydrophobic nano Ni–PTFE electrodeposits. *Wear* **2017**, *374*, 1–4. [CrossRef]

22. Xue, Y.; Wang, S.; Zhao, G.; Taleb, A.; Jin, Y. Fabrication of Ni–Co coating by electrochemical deposition with high super-hydrophobic properties for corrosion protection. *Surf. Coat. Technol.* **2019**, *363*, 352–361. [CrossRef]

23. Tian, X.; Verho, T.; Ras, R.H.A. Moving superhydrophobic surfaces toward real-world applications. *Science* **2016**, *352*, 142–143. [CrossRef] [PubMed]

24. Chung, C.K.; Chang, W.T. Effect of pulse frequency and current density on anomalous composition and nanomechanical property of electrodeposited Ni–Co films. *Thin Solid Films* **2009**, *517*, 4800–4804. [CrossRef]

25. Maksimović, V.M.; Lačnjevac, U.Č.; Stoiljković, M.M.; Pavlović, M.G.; Jović, V.D. Morphology and composition of Ni–Co electrodeposited powders. *Mater. Charact.* **2011**, *62*, 1173–1179. [CrossRef]

26. Yang, Y.; Cheng, Y.F. Electrolytic deposition of Ni–Co–SiC nano-coating for erosion-enhanced corrosion of carbon steel pipes in oilsand slurry. *Surf. Coat. Technol.* **2011**, *205*, 3198–3204. [CrossRef]

27. Rafailović, L.D.; Minić, D.M.; Karnthaler, H.P.; Wosik, J.; Trišović, T.; Nauer, G.E. Study of the dendritic growth of Ni–Co alloys electrodeposited on Cu substrates. *J. Electrochem. Soc.* **2010**, *157*, D295. [CrossRef]

28. Jain, R.; Pitchumani, R. Fractal model for wettability of rough surfaces. *Langmuir* **2017**, *33*, 7181–7190. [CrossRef]

29. *ASM Handbook: Volume 3: Alloy Phase Diagrams*, 10th ed.; ASM International: Novelty, OH, USA, 1992.

30. Wang, N.; Hang, T.; Shanmugam, S.; Li, M. Preparation and characterization of nickel–cobalt alloy nanostructures array fabricated by electrodeposition. *CrystEngComm* **2014**, *16*, 6937–6943. [CrossRef]

31. Zhang, B.; Zhu, Q.; Li, Y.; Hou, B. Facile fluorine-free one step fabrication of superhydrophobic aluminum surface towards self-cleaning and marine anticorrosion. *Chem. Eng. J.* **2018**, *352*, 625–633. [CrossRef]

32. Wang, P.; Li, T.; Zhang, D. Fabrication of non-wetting surfaces on zinc surface as corrosion barrier. *Corros. Sci.* **2017**, *128*, 110–119. [CrossRef]

33. Li, W.; Kang, Z. Fabrication of corrosion resistant superhydrophobic surface with self-cleaning property on magnesium alloy and its mechanical stability. *Surf. Coat. Technol.* **2014**, *253*, 205–213. [CrossRef]

34. Tan, C.; Li, Q.; Cai, P.; Yang, N.; Xi, Z. Fabrication of color-controllable superhydrophobic copper compound coating with decoration performance. *Appl. Surf. Sci.* **2015**, *328*, 623–631. [CrossRef]

35. Qing, Y.; Yang, C.; Yu, Z.; Zhang, Z.; Hu, Q.; Liu, C. Large-area fabrication of superhydrophobic zinc surface with reversible wettability switching and anticorrosion. *J. Electrochem. Soc.* **2016**, *163*, D385–D391. [CrossRef]

coatings

MDPI

Article

Oscillating Magnetic Drop: How to Grade Water-Repellent Surfaces

Angelica Goncalves Dos Santos [1], Francisco Javier Montes-Ruiz Cabello [2], Fernando Vereda [2], Miguel A. Cabrerizo-Vilchez [2] and Miguel A. Rodriguez-Valverde [2,*]

[1] Department of Physics, Florida State University, Tallahassee, FL 32306, USA; amg16k@my.fsu.edu
[2] Biocolloid and Fluid Physics Group, Applied Physics Department, Faculty of Sciences, University of Granada, 18071 Granada, Spain; fjmontes@ugr.es (F.J.M.-R.C.); fvereda@ugr.es (F.V.); mcabre@ugr.es (M.A.C.-V.)
*   Correspondence: marodri@ugr.es; Tel.: +34-958-243-229

Received: 19 February 2019; Accepted: 17 April 2019; Published: 21 April 2019

Abstract: Evaluation of superhydrophobic (SH) surfaces based on contact angle measurements is challenging due to the high mobility of drops and the resolution limits of optical goniometry. For this reason, some alternatives to drop-shape methods have been proposed such as the damped-oscillatory motion of ferrofluid sessile drops produced by an external magnetic field. This approach provides information on surface friction (lateral/shear adhesion) from the kinetic energy dissipation of the drop. In this work, we used this method to compare the low adhesion of four commercial SH coatings (Neverwet, WX2100, Ultraever dry, Hydrobead) formed on glass substrates. As ferrofluid, we used a maghemite aqueous suspension (2% v/v) synthesized ad hoc. The rolling magnetic drop is used as a probe to explore shear solid–liquid adhesion. Additionally, drop energy dissipates due to velocity-dependent viscous stresses developed close to the solid–liquid interface. By fitting the damped harmonic oscillations, we estimated the decay time on each coating. The SH coatings were statistically different by using the mean damping time. The differences found between SH coatings could be ascribed to surface–drop adhesion (contact angle hysteresis and apparent contact area). By using this methodology, we were able to grade meaningfully the liquid-repelling properties of superhydrophobic surfaces.

Keywords: water-repellent surfaces; ferrofluid drop; magnetic field; damped harmonic oscillation

---

1. Introduction

Liquid-repellent surfaces are identified as surfaces with low contact angle hysteresis (<10°) and high contact angles (>150°) [1]. Hysteresis is directly related to the energy cost during the total or partial detachment of a drop from a solid surface. Preparation of superhydrophobic (SH) surfaces is well-established and their water repelling property is commonly evaluated with contact angle or critical sliding angle measurements by using optical goniometry, as happens with the tilting plate method (inclined sessile drop) [2]. However, although this method is useful to illustrate water repellency, it provides low-resolution values of contact angle or critical sliding angle for SH surfaces. The difficult localization of the contact points of non-wetting drops, the insufficient resolution for high contact angles with both the conventional optical devices and numerical fitting of drop profiles [3], the resolution of standard inclinometers working at very low tilt angles (<5°) and the monitoring of "restless" drops placed on SH surfaces [4] required to establish new methodologies. Since the high drop mobility observed on a surface reveals its liquid repellency, the kinetic energy dissipation of a moving sessile drop might quantify the surface friction due to adhesion hysteresis.

A magnetic drop can be manipulated with an external magnetic field (magneto-wetting) [5–11]. The damped-oscillatory motion of a water-based ferrofluid sessile drop driven by a fixed permanent magnet [12] might be used to evaluate experimentally, without further theoretical treatment, the water repellency of non-magnetizable SH surfaces. It is known that, on SH surfaces, a moving sessile drop really rolls, it does not slide [13]. Far from the contact region, the rolling drop moves in a similar way to a rolling rigid-solid. This way, the bulk effect of viscosity may be ignored and the viscous forces mainly act near the contact area [14]. Moreover, this viscous dissipation is further reduced on SH surfaces where the actual drop contact area is particularly low. This approach is different to the viscous forces considered by Timonen et al. [12]. Otherwise, solid–liquid adhesion friction depends on contact angle hysteresis and contact line length of the drop. In this scenario, the plausible differences found with moving drops on SH surfaces would be exclusive to the surface-drop interaction.

In this work, we compared four commercial SH coatings on glass by using the decay time of an oscillating ferrofluid drop released far from its equilibrium. We found the optimal magnetic field to minimize the drop shape distortion and to reproduce longer oscillating motions. We studied the dependence of the damping time on the surface-drop contact area and the drop volume.

2. Materials and Methods

2.1. Fabrication of Superhydrophobic Glass Samples

We evaluated four commercial superhydrophobic coatings: Neverwet Multi-Surface (RUST-OLEUM, Coventry, UK), WX-2100 (Cytonix LLC, Beltsville, MD, USA), Ultra-Ever Dry (Tap Iberica., Burjassot, Spain), and Hydrobead (Hydrobead, San Diego, CA, USA). They were sprayed on clean glass slides, as each supplier recommended. We assumed that these coatings are organic, without metal traces.

2.2. Ferrofluid Preparation

The aqueous ferrofluid was prepared at 2% *v/v* as described elsewhere [15,16]. The process starts with the synthesis of the magnetite nanoparticles by means of the well-known coprecipitation method. These particles are subsequently oxidized to maghemite (γ-Fe_2O_3) with $Fe(NO_3)_3$ and then functionalized with citrate. The electric charge of citrate carboxyl groups at neutral pH prevents particle aggregation. In addition, the small size of the particles (typically 10 nm) in combination to the thermal agitation make the dispersion sedimentation unlikely. As a result, the ferrofluid remains stable for months. The values of density and surface tension of the ferrofluid are 1.04 g/mL and 67.4 mN/m, respectively, which are close to those of pure water. The surface tension indicates that the magnetic nanoparticles have no significant interfacial activity.

2.3. Contact Angle Measurements

The contact angle measurements were conducted with the tilting plate method [2]. We used 100 μL drops of Milli-Q water to increase the density of metastable drop configurations separated by smaller energy barriers and the spatial resolution of the method. Drops were gently deposited at the center of the sample, which is fixed to the tilting platform, oriented horizontally. The drop placement was non-trivial because the drops rolled off the samples very easily. Once the drop was deposited and static, the platform was automatically inclined at a constant rate (5°/s). Side views of the drop were captured simultaneously at 16 fps. We measured the Advancing Contact Angle (ACA) and Receding Contact Angle values (RCA), at both sides of the profile of the deformed drop by using independent elliptical fittings [2]. From these values, we calculated the Contact Angle Hysteresis (CAH) as the difference ACA-RCA.

2.4. Oscillating Magnetic Drop Set-up

The set-up for the oscillating magnetic drop experiments is illustrated in Figure 1. Below the surface, a cylindrical NdFeB magnet with 1.20 cm-diameter and 0.53 cm-height (Supermagnete, Gottmadingen, Germany) was assembled to a vertical aluminum screw. This screw allowed for the adjustment of the surface-to-magnet distance. The magnetic field strength was measured with a teslameter (5170 Gauss/Tesla Meter, FW BELL-EuroMC, Stains, France). Once the SH coating was placed on the stage, the system was leveled out to ensure that the magnetic drop moved randomly on the surface in the absence of the external magnetic field. We deposited three ferrofluid drops with a handheld micropipette on each SH coating. We studied two volumes: 5 and 10 µL. Each drop was initially static thanks to a secondary weak magnet (0.44 cm-height) placed below the surface, at a horizontal distance of 1.76 cm from the primary magnet. The action of this secondary magnet was allowed from an adjustable stop. Once the secondary magnet was moved down, the ferrofluid drop was released, describing a damped oscillatory motion around the primary magnet. During the overall motion, we acquired side views (512 × 128 pixel, 212 pixel/cm) of the back-illuminated drop with a high-speed camera (Phantom, Miro) at 1000 fps. The geometrical drop parameters, such as centroid, contact angles, and contact radius, were calculated by elliptical fitting of each drop contour. The value of contact angle at each side of the drop contour was calculated from the corresponding slope of the best fit evaluated at the contact point (initial and final points of contour). Our resolution was enough to find meaningfully the horizontal positions of drop centroid. We measured the decay time τ of drop motion by fitting the centroid positions to a damped sine wave function ($A\exp(-t/\tau)\sin(\omega t + \varphi)$). We discarded the initial and final oscillations as suggested by Timonen et al. [12].

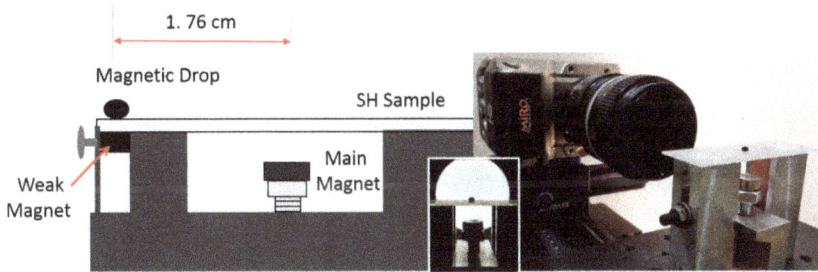

Figure 1. Set-up for oscillating magnetic drop experiments.

3. Results

3.1. Contact Angle Measurements

In Table 1, we show the ACA, RCA, and CAH values measured with the tilting plate method (100 µL-water drops) for the coatings used in this study. These values were averaged over, at least, three experiments. We identify all the coatings as superhydrophobic but no significant difference was found between them.

Table 1. Contact angles of Milli-Q water drops (100 µL) measured with the tilting plane method on the Superhydrophobic (SH) coatings. ACA: Advancing Contact Angle, RCA: Receding Contact Angle and CAH: Contact Angle Hysteresis.

Coating	ACA (°)	RCA (°)	CAH (°)
WX-2100	149 ± 2	147 ± 4	2 ± 6
Hydrobead	152 ± 1	146 ± 3	6 ± 4
Ultra-Ever-Dry	148 ± 2	146 ± 2	2 ± 4
Neverwet	151 ± 3	148 ± 3	3 ± 6

3.2. Magnetic Field Strength

We evaluated the magnetic field strength on the sample stage. This field was maximum close to the primary magnet axis, as expected. We varied the magnet-to-surface distance (r) and the results are shown in Figure 2a. The (maximum) magnetic field strength scales as $1/r$, as predicted by theory.

We also measured the magnetic field strength at different distances from the primary magnet axis (*d*) (Figure 2b), for four surface-to-magnet distances.

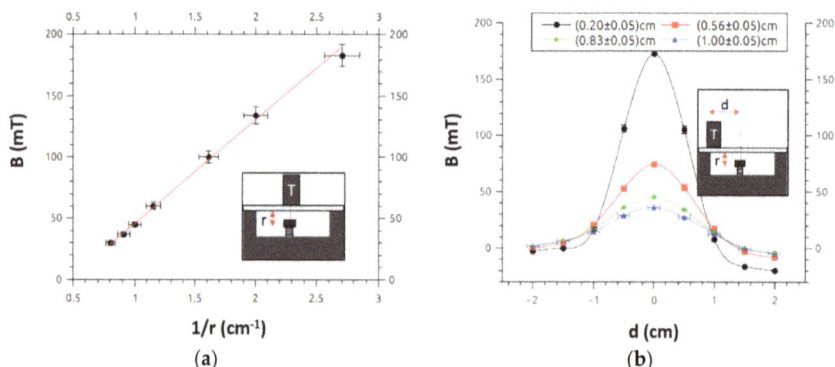

Figure 2. (a) Maximum magnetic field strength (measured at the magnet axis) as a function of the inverse of the surface-to-magnet distance. (b) Magnetic field strength at different horizontal positions on the sample stage from the magnet axis. The symbol "T" in the insets stands for the position of the Teslameter probe.

3.3. Drop Shape Distortion Induced by the Magnetic Field

Magnetic force depends on the field gradient and the ferrofluid magnetization depends on the field strength. Besides, it is known that (surface) magnetic forces alter the wetting response of magnetic drops on solids such as electric forces in electrowetting. The magnetic sessile drop is pressed against the surface by the effect of the external magnetic field, which increases the apparent wet area. We intended to find an optimal surface-to-magnet separation that balances the minimum drop shape distortion and at the same time produces a significantly damped-oscillating drop motion. In particular, we analyzed how the values of drop contact area and contact angle, averaged during the entire drop motion, were modified by the magnetic field. We monitored these parameters for oscillating 5 µL-drops over the Hydrobead coating. In Figure 3a, we show how the inverse of the average drop contact area (1/<*A*>) scales as the surface-to-magnet distance (*r*). From Figure 2a, one may conclude that the contact area increases linearly with the magnetic field strength. In Figure 3b, we show how the average contact angle <*θ*> decreases linearly with the average contact area (<*A*>), as surface-to-magnet distance becomes smaller. The magnetic drop is squashed against the surface as the magnetic force (normal net force) increases, and this increases noticeably the contact area and decreases the contact angle. Short surface-to-magnet distances might produce a total or partial transition in the drop [10] from the Cassie regime (heterogeneous wetting) to the Wenzel regime (homogeneous wetting). In our experiments, for a fixed surface-to-magnet distance, the drop contact area also oscillates because the external magnetic field is not uniform on the SH coating: a greater contact area was observed close to the equilibrium position. This effect complicates the overall drop motion, with a variable period (not pure harmonic). We found that 0.96 cm was the optimal surface-to-magnet distance (peak field of 50 mT) to reproduce a significantly damped-oscillating motion of almost undistorted drops.

We fixed the surface-to-magnet distance to 0.96 cm to evaluate the shape variations of oscillating drops over the SH coatings. In Figure 4, we show the results for two representative coatings (Hydrobead and Neverwet). We plot the average dynamic contact angle (estimated by averaging the contact angles measured at both sides of the drop profile) in terms of the instantaneous contact area, during the complete oscillating motion of three magnetic drops. This plot does not illustrate the response of contact angle hysteresis for each coating. A sessile drop in the Cassie-Baxter regime (on SH surfaces) may reach different configurations (different penetration depths into the asperities) according to the stability of each drop configuration against an external body force (size-dependent). In Table 2, we collect the values of oscillating contact area averaged during the entire drop motion on each coating. Except for Neverwet, the values of mean contact area were very similar. It is also remarkable the significant disagreement for this coating between the contact angles measured with the oscillating drop and the tilting plate method (see Table 1). This behavior will be confirmed in Section 3.4.

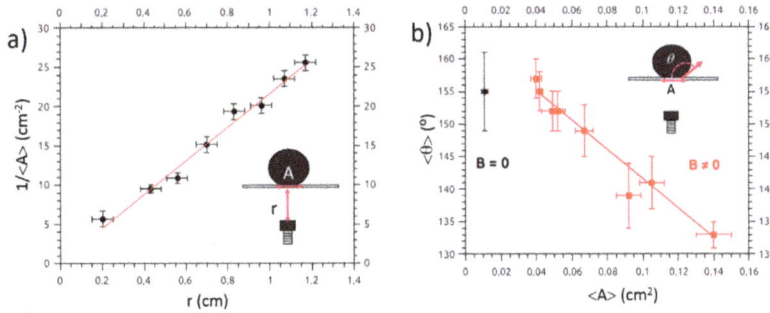

Figure 3. Shape distortion induced by the external magnetic field on an oscillating drop over a Hydrobead coating: (**a**) inverse of the average contact area versus the surface-to-magnet distance; and (**b**) the average contact angle as a function of the average contact area.

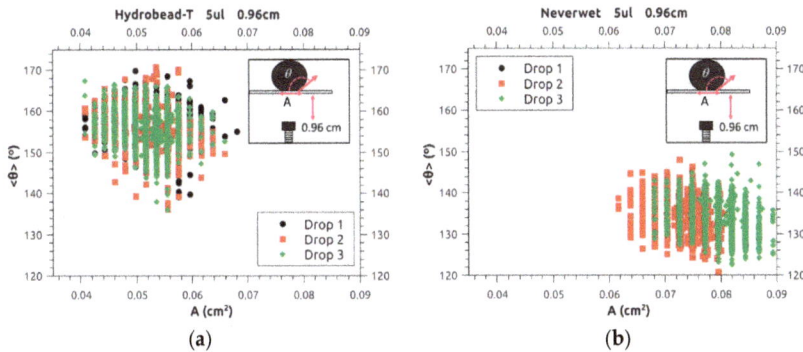

Figure 4. Average dynamic contact angle (measured at both sides of the drop) in terms of the instantaneous contact area for 5 μL-magnetic drops oscillating over glass slides coated with Hydrobead (**a**) and Neverwet (**b**). Each parameter was determined by image analysis of the single frames captured during the complete drop motion.

Table 2. Values of the mean contact area ($<A>$) during the complete drop motion (5 μL) on the SH coatings.

Coating	$<A>$ (mm^2)
WX-2100	5.7 ± 0.4
Hydrobead	5.2 ± 0.4
Ultra-Ever-Dry	5.1 ± 0.4
Neverwet	7.8 ± 0.4

3.4. Dynamics of Oscillating Magnetic Drops

With the surface-to-magnet distance fixed to 0.96 cm, we performed dynamic experiments based on the analysis of the motion of 5 μL-magnetic drops over each SH coating. In Figure 5, we show the evolution of the horizontal position of the drop centroid with time. In all cases, once the drop is released, it oscillates around the equilibrium position but following an underdamped motion with a decay time different for each coating. The horizontal magnetic force acts like a restorative force for horizontal distances below 1 cm respect to the equilibrium position (see Figure 2b). A moving drop on a SH surface describes a superposition of a solid rotation (producing no bulk dissipation) with a viscous friction localized in the contact area. The viscous dissipation is mainly governed by internal flows near contact area rather than in bulk. However, the energy dissipation is also caused by adhesion hysteresis. We reasonably assume that the rolling drops in our experiments undergo a synergetic dissipation due to the viscous stresses developed close to the solid–liquid interface (further reduced in SH surfaces) and the shear adhesion hysteresis. We postulate that a single exponential law, as fit model for the amplitude decay of oscillating drops, enables the capture of the dissipative effects of irreversible solid–liquid adhesion as well as velocity-dependent friction. The goodness of fit was appropriate.

We fitted the horizontal position of the drop centroid (Figure 5) to a damped-harmonic function to determine the damping time (τ). We repeated the experiments with drops of 10 µL and the results are shown in Table 3. In a simplistic scenario of velocity-dependent friction, we would expect greater drop inertia and greater damping. However, this is only found for the Neverwet coating (larger contact area). We found that the Ultra-Ever-Dry coating is the most water repellent (higher damping time), while the WX-2100 coating has the lowest repelling property (lower damping time).

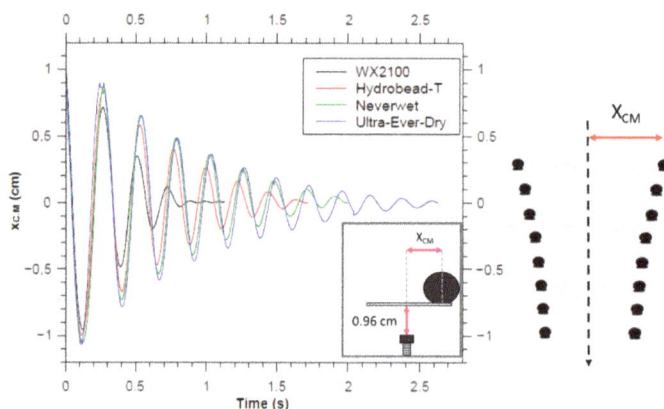

Figure 5. Damped oscillating motion of 5 µL-magnetic drops on the SH coatings, for a surface-to-magnet distance of 0.96 cm. The X_{CM} corresponds to the horizontal deviation of the drop centroid from the equilibrium (primary magnet position).

Table 3. Damping time values for oscillating magnetic drops of 5 and 10 µL on the SH coatings.

Coating	τ (s)-5 µL	τ (s)-10 µL
WX-2100	0.25 ± 0.04	0.40 ± 0.03
Hydrobead	0.61 ± 0.05	0.61 ± 0.06
Ultra-Ever-Dry	0.70 ± 0.10	0.84 ± 0.10
Neverwet	0.66 ± 0.04	0.52 ± 0.09

Like in a typical damper, the frictional force should be a function of the average contact area, but the dependence of the damping time on the size of the contact line is still unclear. We intend to explore the plausible effect of drop contact area on the damping time. In Figure 6a, we plot the inverse of the damping time of 5 µL-magnetic drops on the Hydrobead coating versus the average contact area (modified by varying the surface-to-magnet distance). Below values of 0.1 cm^2 (low magnetic fields), the inverse of the damping time scales linearly with the average contact area of the drop. This is expected because contact angle hysteresis on rough composite surfaces depends on the fractional solid-liquid contact area [17]. However, above 0.1 cm^2, the damping time saturates due to the squashing effect on the magnetic drop, and the possible occurrence of drop configurations far from the 'fakir' state (maximum volume of air entrapped below the drop). In addition to the different wetting properties, the slope of the linear part of Figure 6a (amplified in Figure 6b) may be related to how the water-repelling properties (Cassie regime) of each coating are preserved as the magnetic drop is pressed against the surface. The drops may penetrate sufficiently into the particular surface asperities. Lower values of $1/(\tau<A>)$ for a fixed drop volume would point out to more stable configurations of drop within a well-established Cassie regime because the damping time would be less sensitive to wet area changes. In Table 4, we show the values of the slope $1/(\tau<A>)$ for each coating. This analysis was carried out for drops of 5 and 10 μL. Larger drops typically attain a more stable Cassie configuration. Instead, we observed that the values of $1/(\tau<A>)$ for the Neverwet coating were very similar for both drop volumes. The magnetic force nearly altered the incomplete hybrid wetting regime developed in this coating. In the rest of coatings, the Ultra-Ever-Dry coating was the most stable. This suggest that the Cassie regimen reproduced would be more robust, independent of the drop size.

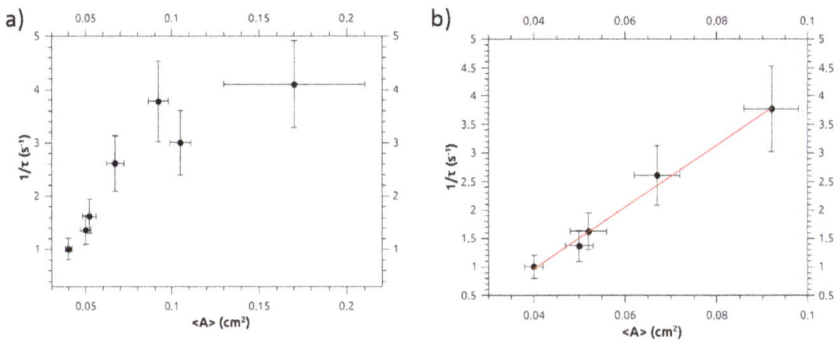

Figure 6. (**a**) Inverse of the damping time $(1/\tau)$ of 5 μL-magnetic drops versus the average contact area $(<A>)$, changed through the external magnetic field, on the Hydrobead coating. (**b**) Linear fit to the section of the same curve corresponding to low average areas (small fields, large surface-to-magnet distances).

Table 4. Values of $1/(\tau<A>)$ $(cm^{-2}\ s^{-1})$ corresponding to 5 and 10 μL for the SH coatings.

Coating	$1/(\tau<A>)$ $(cm^{-2}\ s^{-1})$-5 μL	$1/(\tau<A>)$ $(cm^{-2}\ s^{-1})$-10 μL
WX-2100	70 ± 16	33 ± 7
Hydrobead	32 ± 5	23 ± 9
Ultra-Ever-Dry	28 ± 6	17 ± 3
Neverwet	19 ± 2	19 ± 5

4. Discussion

Direct contact angle and hysteresis measurements of large drops (100 μL) showed that all the coatings were superhydrophobic, but this methodology was unable to identify the more efficient coating. Furthermore, the mapping of average contact angle of small drops (5 μL) in motion in terms of their contact area evidenced the contact angle hysteresis of each surface, but it did not resolve the corresponding water-repelling property. Instead, we were able to grade meaningfully non-magnetizable superhydrophobic surfaces by using the damping time of oscillating magnetic drops ruled by the solid–liquid shear adhesion. We confirmed that this method is more sensitive than goniometry-based methods to validate superhydrophobic surfaces. We recommend using small drops (5–10 μL) of dilute aqueous ferrofluids (2% *v/v*). Moreover, an intense magnetic field (peak value of 50 mT) is recommended to produce a damped oscillating motion with minor changes in the average contact area and contact angle of the drop as compared to the magnetic field-off case. Further work should be addressed to explore the relationship between the damping time of water-repellent surfaces and their contact angle hysteresis, measured with force-based techniques.

Author Contributions: Conceptualization, M.A.R.V. and M.A.C.-V; Methodology, A.G.D.S., F.V. and F.J.M.-R.C.; Writing—Original Draft Preparation, M.A.R.V. and F.J.M.-R.C.; Writing—Review and Editing, M.A.R.V. and F.J.M.-R.C.; Supervision, M.A.R.V. and M.A.C.-V.; Project Administration, M.A.R.V and M.A.C.-V.; Funding Acquisition, M.R.V. and M.A.C.-V.

Funding: This research was financed by the State Research Agency (SRA) and European Regional Development Fund (ERDF) through the project MAT2017-82182-R. Fernando Vereda acknowledges financial support from MAT 2016-78778-R and PCIN-2015-051 projects (Spain).

Acknowledgments: We acknowledge Juan de Vicente (UGR) for the assistance in the measurements of magnetic field strength with teslameter.

Conflicts of Interest: The authors declare no conflict of interest.

References

1. Lv, X.; Tian, D.; Peng, Y.; Li, J.; Jiang, G. Superhydrophobic magnetic reduced graphene oxide-decorated foam for efficient and repeatable oil-water separation. *Appl. Surf. Sci.* **2019**, *466*, 937–945. [CrossRef]
2. Ruiz-Cabello, F.J.M.; Rodriguez-Valverde, M.A.; Cabrerizo-Vilchez, M. A new method for evaluating the most stable contact angle using tilting plate experiments. *Soft Matter* **2011**, *7*, 10457–10461. [CrossRef]
3. Srinivasan, S.; McKinley, G.H.; Cohen, R.E. Assessing the accuracy of contact angle measurements for sessile drops on liquid-repellent surfaces. *Langmuir* **2011**, *27*, 13582–13589. [CrossRef] [PubMed]
4. Tian, X.; Verho, T.; Ras, R.H.A. Moving superhydrophobic surfaces toward real-world applications. *Science* **2016**, *352*, 142–143. [CrossRef] [PubMed]
5. Bormashenko, E.; Pogreb, R.; Bormashenko, Y.; Musin, A.; Stein, T. New investigations on ferrofluidics: Ferrofluidic marbles and magnetic-field-driven drops on superhydrophobic surfaces. *Langmuir* **2008**, *24*, 12119–12122. [CrossRef] [PubMed]
6. Nguyen, N.-T.; Zhu, G.; Chua, Y.-C.; Phan, V.-N.; Tan, S.-H. Magnetowetting and sliding motion of a sessile ferrofluid droplet in the presence of a permanent magnet. *Langmuir* **2010**, *26*, 12553–12559. [CrossRef] [PubMed]
7. Zhou, Q.; Ristenpart, W.D.; Stroeve, P. Magnetically induced decrease in droplet contact angle on nanostructured surfaces. *Langmuir* **2011**, *27*, 11747–11751. [CrossRef] [PubMed]
8. Zhu, G.-P.; Nguyen, N.-T.; Ramanujan, R.V.; Huang, X.-Y. Nonlinear deformation of a ferrofluid droplet in a uniform magnetic field. *Langmuir* **2011**, *27*, 14834–14841. [CrossRef] [PubMed]
9. Cheng, Z.; Lai, H.; Zhang, N.; Sun, K.; Jiang, L. Magnetically induced reversible transition between cassie and wenzel states of superparamagnetic microdroplets on highly hydrophobic silicon surface. *J. Phys. Chem. C* **2012**, *116*, 18796–18802. [CrossRef]
10. Nguyen, N.-T. Deformation of ferrofluid marbles in the presence of a permanent magnet. *Langmuir* **2013**, *29*, 13982–13989. [CrossRef] [PubMed]
11. Manukyan, S.; Schneider, M. Experimental investigation of wetting with magnetic fluids. *Langmuir* **2016**, *32*, 5135–5140. [CrossRef] [PubMed]
12. Timonen, J.V.I.; Latikka, M.; Ikkala, O.; Ras, R.H.A. Free-decay and resonant methods for investigating the fundamental limit of superhydrophobicity. *Nat. Commun.* **2013**, *4*, 2398. [CrossRef] [PubMed]
13. Schellenberger, F.; Encinas, N.; Vollmer, D.; Butt, H.-J. How water advances on superhydrophobic surfaces. *Phys. Rev. Lett.* **2016**, *116*, 096101. [CrossRef] [PubMed]
14. Mahadevan, L.; Pomeau, Y. Rolling droplets. *Phys. Fluids* **1999**, *11*, 2449–2453. [CrossRef]
15. Ménager, C.; Sandre, O.; Mangili, J.; Cabuil, V. Preparation and swelling of hydrophilic magnetic microgels. *Polymer* **2004**, *45*, 2475–2481. [CrossRef]
16. Aldana, S.; Vereda, F.; Hidalgo-Alvarez, R.; de Vicente, J. Facile synthesis of magnetic agarose microfibers by directed self-assembly in W/O emulsions. *Polymer* **2016**, *93*, 61–64. [CrossRef]
17. Nosonovsky, M. Model for solid-liquid and solid-solid friction of rough surfaces with adhesion hysteresis. *J. Chem. Phys.* **2007**, *126*, 224701. [CrossRef] [PubMed]

Coatings **2019**, *9*, 270

coatings

MDPI

Article

Surfactant-Free Electroless Codeposition of Ni–P–MoS$_2$/Al$_2$O$_3$ Composite Coatings

Ping Liu [1,2,3,4,5,*], **Yongwei Zhu** [1], **Qi Shen** [1], **Meifu Jin** [2], **Gaoyan Zhong** [2], **Zhiwei Hou** [4],
Xiao Zhao [5], **Shuncai Wang** [5] **and Shoufeng Yang** [5,6]

[1]   Jiangsu Key Laboratory of Precision and Micro-Manufacturing Technology, Nanjing University of
    Aeronautics and Astronautics, Nanjing 210016, China; meeywzhu@nuaa.edu.cn (Y.Z.);
    sevenshenqi@163.com (Q.S.)
[2]   College of Engineering, Nanjing Agricultural University, Nanjing 210031, China;
    jinmeifu@njau.edu.cn (M.J.); gyzhong@njau.edu.cn (G.Z.)
[3]   Key Laboratory of Modern Agricultural Equipment and Technology, Ministry of Education/High-tech Key
    Laboratory of Agricultural Equipment & Intelligentization of Jiangsu Province, Jiangsu University,
    Zhenjiang 212013, China
[4]   Jiangsu Key Laboratory of Advanced Manufacturing Technology, Huaiyin Institute of Technology,
    Nanjing 223003, China; zw_hou66@163.com
[5]   Faculty of Engineering and Environment, University of Southampton, Southampton SO17 1BJ, UK;
    xiao.zhao@soton.ac.uk (X.Z.); wangs@soton.ac.uk (S.W.); s.yang@soton.ac.uk (S.Y.)
[6]   Production Engineering, Machine Design and Automation, Department of Mechanical Engineering,
    Katholieke Universiteit Leuven (KU Leuven), Leuven 3001, Belgium
*   Correspondence: liuping@njau.edu.cn; Tel.: +86-25-5860-6603

Received: 20 December 2018; Accepted: 10 February 2019; Published: 13 February 2019

Abstract: This paper presents the influence of an inorganic Al$_2$O$_3$ layer over MoS$_2$ particles on the tribological performance of electroless Ni–P–MoS$_2$/Al$_2$O$_3$ composite coatings fabricated without using surfactants. The Al$_2$O$_3$-coated MoS$_2$ particles were prepared by a heterogeneous nucleation process. The dry sliding tests of the composite coatings were tested against a WC ball. SEM was used to observe the surface morphology of particles, composite coatings, and worn surfaces. The results indicate that the coverage of an Al$_2$O$_3$ coating on MoS$_2$ particles significantly affects the surface morphology, frictional coefficient and wear loss of the composite coatings. The incorporation of Al$_2$O$_3$-coated MoS$_2$ particles with lower coverage (up to 7% of Al$_2$O$_3$) could obtain compact surface structure of composite coatings, which contribute to reduced wear loss. However, higher coverage would lead to loose surface structure of the composite coatings, and thus increase their wear loss.

Keywords: electroless composite coating; Al$_2$O$_3$-coated particles; MoS$_2$ particles; wear resistance; surfactant

1. Introduction

Electroless Nickel (EN) composite coatings containing submicro/nano-sized particles in a nickel matrix have received increasing attention in recent years [1–3]. The incorporation of solid particles into the matrix could remarkably improve the mechanical and physiochemical properties of composite coatings. For example, hard particles such as Al$_2$O$_3$ [4,5], SiC [6,7], SiO$_2$ [8], TiO$_2$ [9], and diamond [10] enhance the hardness and wear resistance of composite coatings. Solid lubricant particles such as MoS$_2$ [11,12], PTFE [13], and BN(h) [14] lower the frictional coefficient of composite coatings and consequently reduces wear loss. It was found that these superior properties highly rely upon a homogeneous distribution of particles in the coating matrix. However, most of the particles have a strong tendency towards agglomeration in an EN solution [15–18]. To prevent agglomeration, surfactants are in particular added into the EN plating bath [19–21].

Surfactants (or surface active substances) are usually organic compounds that are amphiphilic. These lower the interfacial tension between particles and the EN solution by improving the wettability of particles [22]. These additives are very important in the incorporation of second phase particles, especially for water-repellent ones (e.g., MoS_2, PTFE) [23]. Without them, these hydrophobic particles would not be well immersed in the EN plating solution [22–24].

However, the adverse effects of using surfactants to produce composite coatings have recently been reported. Sudagar et al. [25] stated that the deposition of coating would be delayed (by as much 40 min) at the early stage due to the indirect contact of electrolytes with substrate caused by surfactant coverage. Zielinska et al. [26] discovered the coverage of surfactants on Ni ions and hypophosphite ions reduced the amounts of nickel and phosphorus in coatings by hampering the reduction process of nickel ions. The surfactant coverage on substrates also provided a barrier for the deposition of the coating, and thus decreased the deposition rate of composite coatings [27–31]. Mai et al. [32] argued that the introduction of additives weakens the interfacial bonding of particles and matrix, fading the properties of composite coatings. Furthermore, due to the complexity and selectivity of surfactants, numerous extra steps are required to identify suitable types and concentrations of surfactants for electroless composite plating [33–36]. Unfortunately, it is quite challenging to choose an appropriate surfactant for a specific plating configuration.

Recently, surface modification of inorganic coatings on particles has received considerable attention in several fields [37–39]. Our previous work [40] indicated that Al_2O_3 loading on the particles could improve the wettability of hydrophobic MoS_2 particles. As a result, we successfully incorporated the coated MoS_2 particles into a nickel matrix by the electroless plating method in the absence of surfactants [41–43], as shown in Figure 1. The resultant composite coatings showed improvements in wear property compared to those incorporated uncoated MoS_2 with aids of surfactants. These results indicated that the environmentally hazardous surfactants could be reduced or even excluded. In keeping with the nature of particles, an excessive coverage over particles is not expected. However, there are as yet only a few reports on the influence of particle coverage on composite coating performance.

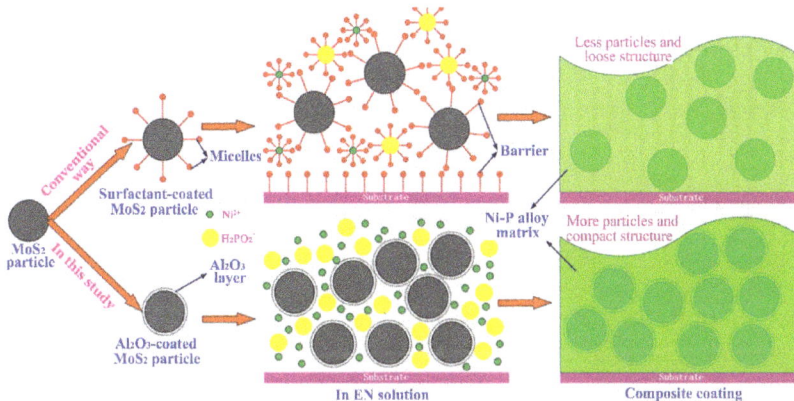

Figure 1. Diagram of the comparison between the codeposition of bare MoS_2 particles with surfactant and Al_2O_3-coated MoS_2 particles without surfactants into an electroless nickel matrix.

This work aims to evaluate the influence of particle coverage on the wear properties of the EN composite coatings. We prepared the Al_2O_3-coated MoS_2 particles with various coverages, and then used them to fabricate electroless composite coatings without using surfactants. The surface morphology, friction coefficient and wear loss of these composite coatings were investigated as well.

2. Materials and Methods

2.1. Preparation of Coated Particles

MoS_2 particles with an average particle size of Φ0.5 μm (supplied by Shanghai Haochem Company, Shanghai, China) were used to prepare the Al_2O_3-coated particles. In this process, the heterogeneous nucleation method was used (the detailed procedure may be found in our previous work [40]). Firstly, MoS_2 powder was etched 30 min with 20 wt %. H_2SO_4 solution at 80 °C to eliminate the oxide surface, and then cleaned with deionized (DI) water. Three grams of MoS_2 powder were added into 300 mL of NaOAc and HOAc-buffered solution with a pH of 4.5 to produce suspension. $Al(OH)_3$-coated MoS_2 particles were obtained by adding, via drops, 0.2 M $Al(NO_3)_3$ solution to the suspension, at an estimated rate of 0.05 mL/s. The reaction temperature was 60 °C. The pH was subjected to the addition of $Al(NO_3)_3$ solution, and kept at 4.5 ± 0.2. The loading of $Al(OH)_3$ on the particles was controlled by the reaction time. The reaction production was dried at 120 °C for 12 h, and subjected to dehydration in air at 350 °C for 2 h to produce Al_2O_3-coated particles. We designed a series of Al_2O_3-coated particles with amounts of coating ranging from 5% to 20% in increments of 5%, and a coating amount of 40% of the layer. However, we finally obtained the samples with loadings of 3%, 7%, 11%, 24% and 42%, respectively. The differences were attributed to the pH variability with the increased total volume of the suspension, which significantly affected the loading amount of $Al(OH)_3$.

The amounts of coating were determined by total water loss rate after being heated from 30 to 1200 °C at the heating rate of 20 K/min under Ar atmosphere condition. The decomposing reaction equation of $Al(OH)_3$ into Al_2O_3 and H_2O reveals that H_2O accounts for 34.6%, Al_2O_3 65.4%. Given the total water loss rate λ, then the coverage of Al_2O_3 can be determined by the following equation (Equation (1)),

$$Al_2O_3 \text{ (wt \%)} = 65.4\%/34.6\% \, \lambda = 1.89 \, \lambda. \tag{1}$$

2.2. Preparation of Ni–P Matrix Composite Coatings

2.2.1. Substrate Preparation

Medium carbon steel specimens with the size of Φ 50 mm × 2 mm were used as substrates. Each substrate was polished using a 2000 grade abrasive paper and ultrasonically cleaned in acetone. All the substrates were degreased with an alkaline solution (Na_2CO_3 30 g/L, NaOH 30 g/L, $Na_3PO_4 \cdot 12H_2O$ 10 g/L, Na_2SiO_3 10 g/L, OP-10 2 mL/L) at 70–80 °C for 60 min, and then cleaned up with deionized water. After that, the substrates were activated in a 20 wt % H_2SO_4 aqueous solution for 90 s and rinsed with deionized water twice prior to plating.

2.2.2. Electroless Plating Bath and Operating Conditions

The commercial electroless plating solution (HK350 from Haibo Co. Ltd., Nanjing, China) was used to produce EN composite coatings. The composition of the plating solution mainly consists of $NiSO_4$ 25 g/L, $NaH_2PO_2 \cdot H_2O$ 22 g/L, buffer agents, and stabilized agents. The plating process took place in a 500 mL thermostated vessel. For comparison, Ni–P–MoS_2 composite coating was fabricated as well. The pre-treated particles were first added to a separate portion of EN solution and dispersed by an ultrasonic cleaner for 20 min. In this step, it is easy to produce MoS_2 suspension for the coated particles without the aids of surfactant (see Figure 2a). However, the uncoated ones must use the surfactant of Cetyltrimethylammonium Bromide (CTAB) for this purpose (see Figure 2b). Then, the suspension was transferred to the main EN solution to produce a composite plating bath with a concentration of 1 g/L of particles. All the samples were pre-plated with an active layer of Ni–P alloy at 88 °C for 10 min before applying the composite coating plating. After that, the samples were placed vertically into the composite plating bath under these conditions: A pH of 5.0, a magnetic stirring rate of 700 rpm, and a duration of 90 min at 88 °C.

Figure 2. Particles suspension in EN plating bath of (**a**) Al$_2$O$_3$-coated MoS$_2$ particles, (**b**) pristine MoS$_2$ particles, respectively.

2.3. Post-Treatment of Electroless Coatings

All the as-prepared coatings have a similar coating thickness of 22–25 μm under the operation condition mentioned above. After preparation, these samples were first heated at 200 °C for 1 h to be dehydrogenated, and then heated at 400 °C for 1 h in air to reinforce the Ni–P matrix of composite coating [4].

2.4. SEM/EDS/XRD/LSCM Characterization

SEM (Hitachi S-4800, Tokyo, Japan) was used to observe the surface morphology of the coated MoS$_2$ particles and the composite coatings. EDS (Bruker EDS QUANTAX, Billerica, MA, USA) was employed to analyze the surface chemical composition of the coated particles. XRD (X'Pert PRO, Malvern Panalytical, Almelo, The Netherlands) with Cu Kα radiation was utilized to characterize the phase structure of the coated particles and composite coatings after heat treatment. The Laser-Scanning Confocal Microscope (LSCM, Olympus, OLS4100, Tokyo, Japan) was used to obtain the surface roughness of composite coatings.

2.5. Friction and Wear Tests

The friction and wear tests were carried out on a tribometer (CFT-1, Lanzhou, China) with a pin-on-disc contact configuration under dry sliding conditions. The composite coating specimens were used as rotating discs, and a tungsten carbide (WC) ball with a diameter of 6 mm was used as a fixed counterpart. For all the tests, the sliding velocity was fixed at 0.5 m/s with a contact radius of 12 mm, duration time of 4 h and normal load of 9.3 N. The worn scar was observed by using SEM (Hitachi S-4800) as well.

3. Results and Discussion

3.1. Al$_2$O$_3$-Coated MoS$_2$ Particles

Figure 3 illustrates the typical SEM morphologies of bare MoS$_2$ particles and the Al$_2$O$_3$-coated ones with the coverage of 7%, 24%, and 42%. The morphology of pristine MoS$_2$ particles is flat and smooth [40]. However, the surface of the coated particles became very rough, and the roughness varies with the amounts of Al$_2$O$_3$ loading. A lower amount of coating produces a partial cover on particles while a higher one produces an entire cover. It is worth mentioning that currently it is difficult to obtain a homogeneous coating layer over MoS$_2$ particles due to the strong hydrophobic feature of their pristine surfaces.

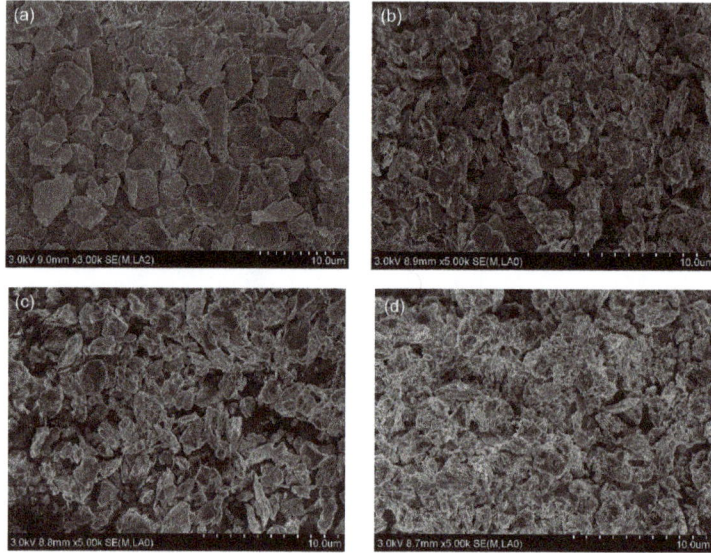

Figure 3. SEM morphology of MoS$_2$ particles (**a**) pristine particles [40], (**b**) Al$_2$O$_3$-coated particles with coverage of 7%, (**c**) Al$_2$O$_3$-coated particles with coverage of 24%, (**d**) Al$_2$O$_3$-coated particles with coverage of 42%, respectively. Adapted with permission from [40]. Copyright 2015 Elsevier.

The chemical compositions of the surface for the coated particles are determined by EDS and the structure is characterized by XRD, as discussed elsewhere [40]. EDS scanning over the surface of the as-coated particles with the coverage of 42% shows the Mo, S, O, and Al element is about 32 wt %, 16 wt %, 43 wt %, and 9 wt %, respectively. This result indicates the formation of aluminum hydroxide on the surface of particles. Figure 4 shows the XRD diffraction pattern of the coated particles with the coverage of 7% and 42% respectively after calcination at 350 °C. The XRD pattern reveals that the surface coating on particles is essentially an amorphous structure. Apart from MoS$_2$, no diffraction peak corresponding to alumina or aluminum hydroxide could be found. For comparison, the XRD diffraction pattern of Al(OH)$_3$ and Al$_2$O$_3$ are presented in our previous work [40]. The main reason is that the calcining temperature of 350 °C is insufficient to crystallize the as-amorphous structure of the aluminum hydroxide [44].

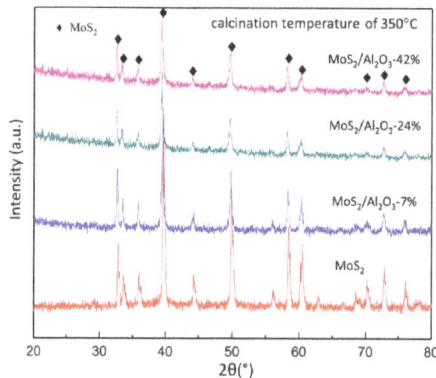

Figure 4. XRD pattern of MoS$_2$ particles and Al$_2$O$_3$-coated MoS$_2$ particles with loading of 7%, 24% and 42%, respectively, after calcining at 350 °C.

3.2. Ni–P Matrix Composite Coatings

Figure 5 shows the XRD patterns of the composite coatings containing Al_2O_3-coated MoS_2 particles with various coverage after heat treatment at 400 °C. The commercial electroless plating solution used in this study produces an amorphous Ni–P alloy matrix according to the indication of the manufacturer although the percentage of P cannot be evaluated. As a result, the transformation of the amorphous Ni–P matrix into Ni and Ni_3P phases took place after the heating treatment as most of the previous works have reported [4,45,46]. The phases transformation is independent of incorporation MoS_2 particles. This study reveals that the transformation is not affected whether the MoS_2 particles coated with Al_2O_3 or not.

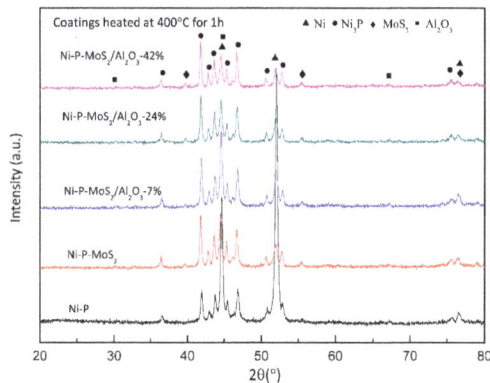

Figure 5. XRD patterns of EN composite coatings containing Al_2O_3-coated MoS_2 particles with various coverage after heat treatment at 400 °C.

The diffraction patterns corresponding to MoS_2 phase on all the composite coatings indicate that the MoS_2 particles have been successfully incorporated into the Ni–P matrix, have not involved in the transformation of the Ni–P matrix when heated. Very weak diffraction peaks of Al_2O_3 can be found for the composite coatings with coated MoS_2. The results are mainly attributed to a further dehydration and crystallization of aluminum hydroxide into alumina after the heating treatment at 400 °C as it exceeds the calcination temperature of 350 °C. However, the very small amount results in the weak diffraction intensity.

Figure 6 indicates the surface morphology and 3D images of EN composite coatings containing the coated MoS_2 particles with various Al_2O_3 coverage after heat treatment. The Ni–P–MoS_2 composite coating prepared using CTAB shows a spherical nodular structure, which is in line with most of the other literature [25,47,48]. The surface roughness S_a is 6.528 μm, as shown in Figure 6(a-2). A similar structure can also be seen on the surface of the composite coating with Al_2O_3 coverage of 3%. However, the latter shows a finer and homogeneous surface, whose surface roughness S_a is 3.135 μm, much less than the former. The increase of the coverage up to 7% could lead to a more compact surface structure of composite coating. The surface roughness decreased to S_a = 1.501 μm, as can be seen in Figure 6(c-2). The reason might be attributed to the role of Al_2O_3 loading on MoS_2 contributing to the fine grain size of the composite coatings. The Al_2O_3 loading could remarkably enhance the wettability of MoS_2 according to our previous findings [40]. This result indicates that less coverage of Al_2O_3 on particles could make it feasible to produce composite coatings without using surfactants.

The further increase of the coverage, however, would induce a reverse change of the surface morphology. The coatings show a loose surface structure with small nodules, large bumps and deep micropores, as shown in Figure 6e,f. The surface roughness also increases from S_a = 2.838 μm to S_a = 9.511 μm with the increase of coverage. This result might be attributed to the split of alumina from MoS_2 particles due to a large difference in elastic modulus between brittle Al_2O_3 and soft MoS_2.

The broken MoS$_2$ tends to form a film on the interface of plating bath, especially in the case of particles with high coverage. The MoS$_2$ film would hamper the escape of the hydrogen produced in the plating reaction, and result in the large bumps and deep micropores of the composite coating. On the other hand, the naked MoS$_2$ might be codeposited into the composite coating, as shown in Figure 6e,f (naked MoS$_2$). The incorporation of MoS$_2$ particles with highest loading of Al$_2$O$_3$ could lead to so very loose coating strucuture that the cross section of it could not obtained. Therefore, the higher coverage on particles is not recommended.

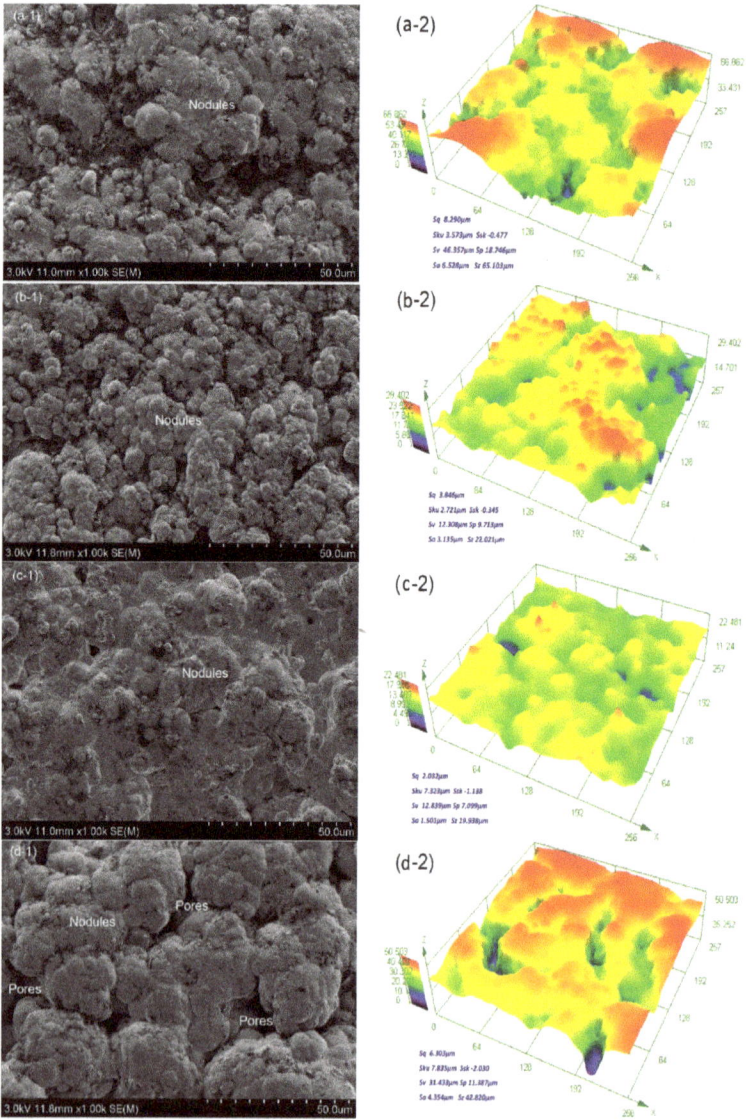

Figure 6. *Cont.*

Figure 6. SEM morphology and 3D images of EN composite coatings containing Al$_2$O$_3$-coated MoS$_2$ particles with various coverage after heat treatment at 400 °C: (**a**) 0%, (**b**) 3%, (**c**) 7%, (**d**) 11%, (**e**) 24%, (**f**) 42%, respectively.

3.3. Friction Coefficient

Figure 7 shows the evolution of friction coefficients for the EN composite coatings incorporating Al$_2$O$_3$-coated MoS$_2$ particles with various coverage while running against WC. Due to the very high hardness of its counterpart (WC), the Ni–P coating has a friction coefficient of approximately 0.6 at steady state after the running-in stage. Compared with the Ni–P alloy, the Ni–P–MoS$_2$ composite coating shows a significant decrease in the friction coefficient due to the lubricant effect of the MoS$_2$. However, its friction coefficient could not reach a steady state until it reached a sliding distance of up to 850 m, which is similar to that in the literature [49]. After that, its frictional coefficient tends to be stable at 0.4. This result is much smaller than that in the literature [11], which might be attributed to the submicro-sized particles used in this study.

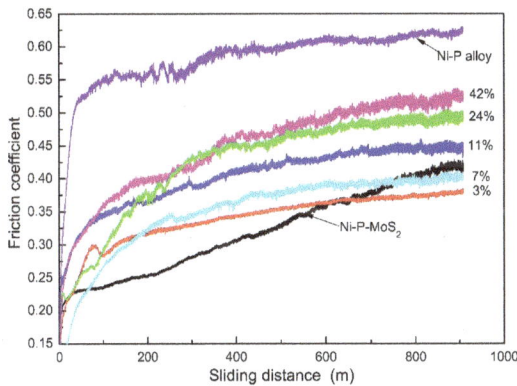

Figure 7. Friction coefficient of electroless composite coatings incorporating Al$_2$O$_3$-coated MoS$_2$ particles with various coverage.

The composite coatings with the coated MoS$_2$ indicate a relative steady evolution in the friction coefficients, which might be attributed to the fabrication without surfactants. The friction coefficients increase with the increase of Al$_2$O$_3$ coverage of the particles accordingly. Higher coverage corresponds to the higher friction coefficients. This result might be caused by the Al$_2$O$_3$ on the particles. The Al$_2$O$_3$ might be involved in friction behavior and lead to an increase in the friction coefficient.

3.4. Wear

Figure 8 demonstrates the mass loss of EN composite coatings incorporating Al$_2$O$_3$-coated MoS$_2$ particles with various coverage after wear. The mass loss of Ni–P–MoS$_2$ composite coating, about 0.63 mg/km, is lower than that of the Ni–P coating, which is approximately 0.70 mg/km. The less mass loss might be resulted from its remarkably decreased friction coefficient. Compared with Ni–P≠MoS$_2$ composite coating, the composite coatings containing the coated MoS$_2$ with the coverage of 3% and 7% show a further reduction in mass loss in the range of 0.52–0.54 mg/km. The result could mainly be attributed to their fine and compact structures, which avoid the side-effects of surfactants. However, the composite coatings with higher coverage of MoS$_2$ show very large mass loss. The one with the coverage of 42% can reach 0.73 mg/km, exceeding the Ni–P alloy. The reason is mainly because it has a deteriorated surface structure and results in weak wear resistance.

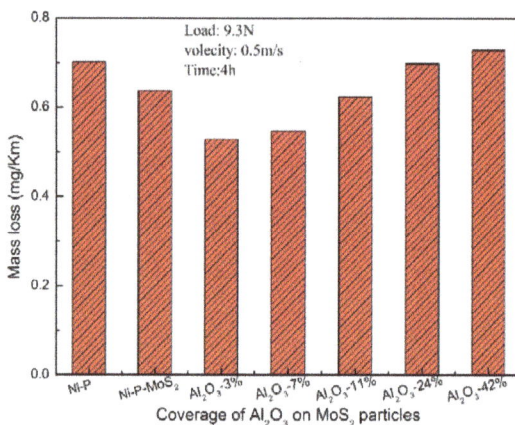

Figure 8. Mass loss of EN composite coatings incorporating Al$_2$O$_3$-coated MoS$_2$ particles with various coverage.

Figure 9 presents SEM of the worn tracks for the composite coatings incorporating the coated MoS$_2$ with various coverage. The Ni–P–MoS$_2$ composite coating shows a rough worn surface with more ploughing and scuffing tracks. This is attributed to its relatively coarse structure, which provides severe wear and thus leads to an increasing friction coefficient when against the hard pair of WC materials, as shown in Figure 7. Both the composite coatings with the coverages of 3% and 7% show a relatively smooth and flat worn surface without obvious ploughing tracks. The results are in accordance with their respective steady evolution of the friction coefficient, and are mainly attributed to their respective fine and compact microstructures. However, the composite coatings containing MoS$_2$ with higher coverage of Al$_2$O$_3$ indicate a remarkably rough worn surface with numerous fine and closed-packed scuffing tracks. One main reason is that the loose coating structure has a lower bearing capacity for shear load during the sliding test. The other reason is that the free Al$_2$O$_3$ stripping from the surface of MoS$_2$ could turn into abrasive particles and accordingly intensify the interface destruction of both parts. As a result, these composite coatings show higher friction coefficients and worse wear resistance.

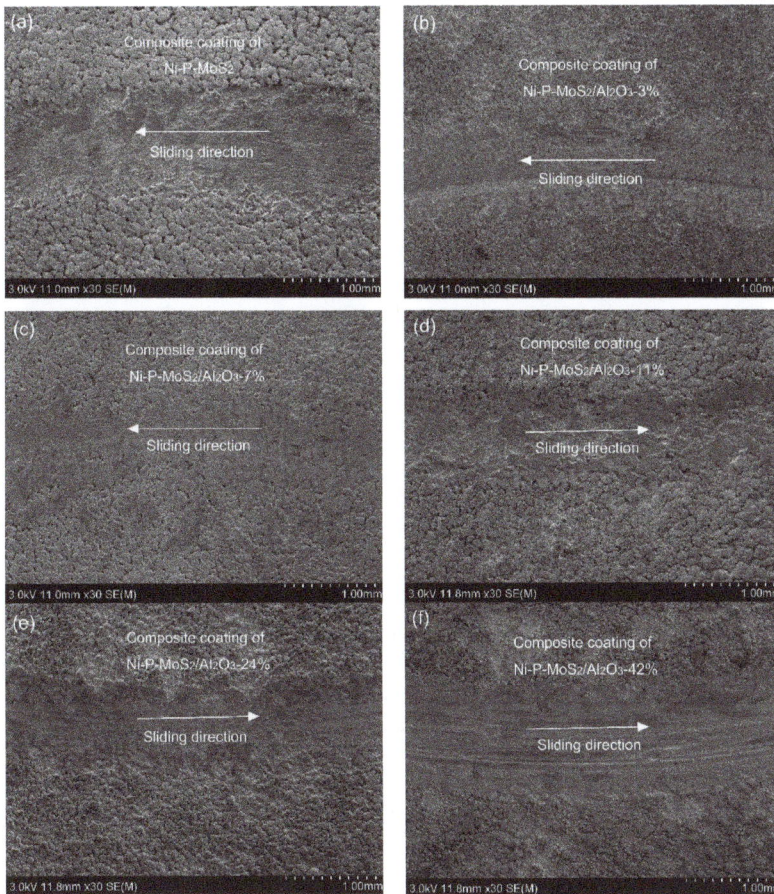

Figure 9. Worn track profile of electroless composite coatings incorporating Al$_2$O$_3$-coated MoS$_2$ particles with various coverages of (**a**) 0%, (**b**) 3%, (**c**) 7%, (**d**) 11%, (**e**) 24%, (**f**) 42%, respectively.

Figure 10 reveals the worn scar of the WC counter after sliding against the electroless Ni–P coating and the composite coatings containing MoS$_2$ with various coverage of Al$_2$O$_3$. The worn scar sliding against Ni–P coating shows a small and regular area with numerous ploughing tracks. Due to the MoS$_2$ lubricant, the worn scar sliding against Ni–P–MoS$_2$ composite coating is relatively smooth. A similar result is shown on the surface of counter-sliding against the composite coating with 3% coverage MoS$_2$ as well, which indicates the lesser coverage of Al$_2$O$_3$ does not change the lubricant property of MoS$_2$. When the coverage is up to 7%, the corresponding worn scar of the counter appears to be slightly rough. The further increase of the coverage of Al$_2$O$_3$ on MoS$_2$ would lead to a rougher worn surface on the counter, sliding against them. The numerous scuffing tracks on the worn scars can be seen. Apart from the loose surface of the composite coating, the stripping free Al$_2$O$_3$ from MoS$_2$ particles might be another reason for the deterioration of the worn surface.

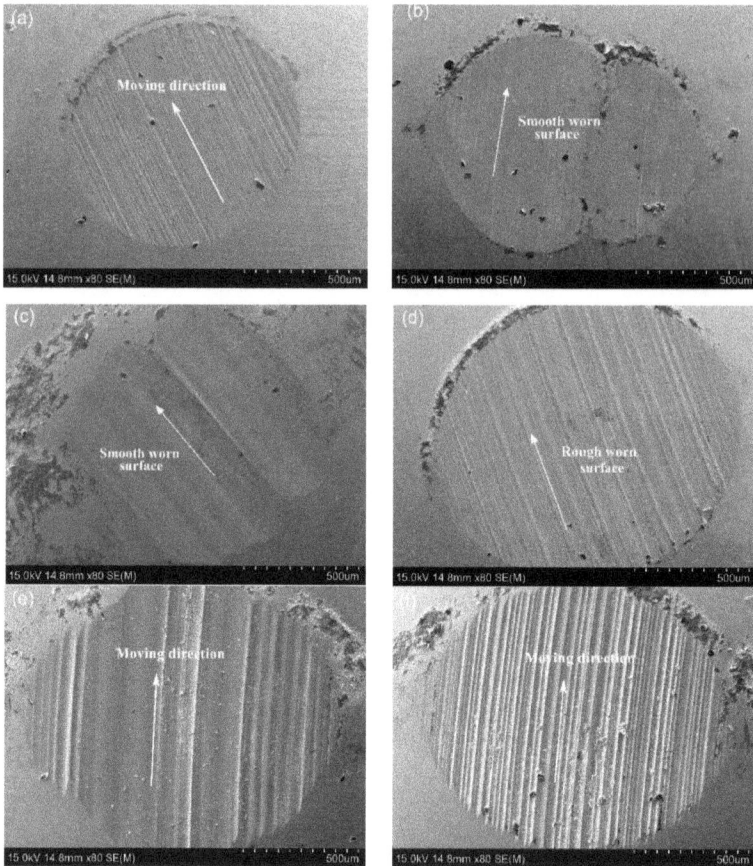

Figure 10. The worn scars of the WC counter after sliding against (**a**) EN, and its composite coatings incorporating Al_2O_3-coated MoS_2 particles with coverage of (**b**) 0%, (**c**) 3%, (**d**) 7%, (**e**) 24%, (**f**) 42%.

4. Conclusions

The Al_2O_3-coated MoS_2 particles with various coverages were obtained by using the heterogeneous nucleation process. The composite coatings with the coated particles were fabricated successfully in the absence of surfactants. The Al_2O_3 loading on MoS_2 particles shows significant influence on the friction and wear performance of composite coatings under dry sliding configuration. The composite coatings containing MoS_2 with lower Al_2O_3 loading show a fine roughness and compact structure, and thus correspond to higher wear resistance. Those containing MoS_2 with higher Al_2O_3 loading show a loose structure, and have less wear resistance. This study reveals that up to 7% Al_2O_3 coverage can achieve a quality composite coating without using surfactants. The small amount of Al_2O_3 offers the advantage of affecting the lubricant nature of MoS_2 particles much less, which in turn improves the wear property of the composite coating.

Author Contributions: Conceptualization, Y.Z.; Methodology, G.Z.; Formal Analysis, Z.H. and X.Z.; Investigation, M.J. and Q.S.; Writing, P.L.; Review and Editing, S.W. and S.Y.

Funding: The work was funded by the National Natural Science Foundation (No. 51675276), the Jiangsu Key Laboratory of Precision and Micro-Manufacturing Technology (Nanjing University of Aeronautics and Astronautics) (No. JSPM201701), the Key Laboratory of Modern Agricultural Equipment and Technology (Jiangsu University), Ministry of Education(MOE)/High-tech Key Laboratory of Agricultural Equipment & Intelligentization of Jiangsu Province (No. NZ201606), and the Jiangsu Key Laboratory of Advanced

Manufacturing Technology (Huaiyin Institute of Technology) (No. HGAMTL-1606). The corresponding author would like to acknowledge the scholarship program of the China Scholarship Council (CSC), MOE, China (No. 20176855025).

Conflicts of Interest: The authors declare no conflict of interest.

References

1. Makkar, P.; Mishra, D.D.; Agarwala, R.C.; Agarwala, V. A novel electroless plating of Ni–P–Al–ZrO$_2$ nanocomposite coatings and their properties. *Ceram. Int.* **2014**, *40*, 12013–12021. [CrossRef]
2. Ardakani, S.R.; Afshar, A.; Sadreddini, S.; Ghanbari, A.A. Characterization of Ni–P–SiO$_2$–Al$_2$O$_3$ nano-composite coatings on aluminum substrate. *Mater. Chem. Phys.* **2017**, *189*, 207–214. [CrossRef]
3. Wu, H.; Liu, F.; Gong, W.; Ye, F.; Hao, L.; Jiang, J.; Han, S. Preparation of Ni–P–GO composite coatings and its mechanical properties. *Surf. Coat. Technol.* **2015**, *272*, 25–32. [CrossRef]
4. Novak, M.; Vojtech, D.; Vitu, T. Influence of heat treatment on tribological properties of electroless Ni–P and Ni-P-Al$_2$O$_3$ coatings on Al-Si casting alloy. *Appl. Surf. Sci.* **2010**, *256*, 2956–2960. [CrossRef]
5. Rahimi, A.R.; Modarress, H.; Iranagh, S.A. Effect of alumina nanoparticles as nanocomposites on morphology and corrosion resistance of electroless Ni–P coatings. *Surf. Eng.* **2011**, *27*, 26–31. [CrossRef]
6. Wang, Q.; Callisti, M.; Greer, J.; McKay, B.; Milickovic, T.K.; Zoikis-Karathanasis, A.; Deligkiozi, I.; Polcar, T. Effect of annealing temperature on microstructure, mechanical and tribological properties of nano-SiC reinforced Ni–P coatings. *Wear* **2016**, *356*, 86–93. [CrossRef]
7. Soleimani, R.; Mahboubi, F.; Kazemi, M.; Arman, S.Y. Corrosion and tribological behavior of electroless Ni–P/nano–SiC composite coating on aluminium 6061. *Surf. Eng.* **2015**, *31*, 714–721. [CrossRef]
8. Sadeghzadeh-Attar, A.; AyubiKia, G.; Ehteshamzadeh, M. Improvement in tribological behavior of novel sol-enhanced electroless Ni–P–SiO$_2$ nanocomposite coatings. *Surf. Coat. Technol.* **2016**, *306*, 837–848. [CrossRef]
9. Hosseini, J.; Bodaghi, A. Corrosion behaviour of electroless Ni–P–TiO$_2$ nanocomposite coatings using Taguchi. *Surf. Eng.* **2013**, *29*, 183–189. [CrossRef]
10. Reddy, V.; Ramamoorthy, B.; Nair, P. A study on the wear resistance of electroless Ni–P/Diamond composite coatings. *Wear* **2000**, *239*, 111–116. [CrossRef]
11. Li, Z.; Wang, J.; Lu, J.; Meng, J. Tribological characteristics of electroless Ni–P–MoS$_2$ composite coatings at elevated temperatures. *Appl. Surf. Sci.* **2013**, *264*, 516–521. [CrossRef]
12. Hu, X.; Jiang, P.; Wan, J.; Xu, Y.; Sun, X. Study of corrosion and friction reduction of electroless N–P coating with molybdenum disulfide nanoparticles. *J. Coat. Technol. Res.* **2009**, *6*, 275–281. [CrossRef]
13. Srinivasan, K.N.; John, S. Studies on electroless nickel–PTFE composite coatings. *Surf. Eng.* **2005**, *21*, 156–160. [CrossRef]
14. Hsua, C.I.; Hou, K.H.; Ger, M.D.; Wang, G.L. The effect of incorporated self-lubricated BN(h) particles on the tribological properties of Ni–P–BN(h) composite coatings. *Appl. Surf. Sci.* **2015**, *357*, 1727–1735. [CrossRef]
15. De Hazan, Y.; Knies, F.; Burnat, D.; Graule, T.; Yamada-Pittini, Y.; Aneziris, C.; Kraak, M. Homogeneous functional Ni–P/ceramic nanocomposite coatings via stable dispersions in electroless nickel electrolytes. *J. Colloid Interface Sci.* **2012**, *365*, 163–171. [CrossRef] [PubMed]
16. Sharma, A.; Singh, A.K. Electroless Ni–P–PTFE-Al$_2$O$_3$ dispersion nanocomposite coating for corrosion and wear resistance. *J. Mater. Eng. Perform.* **2014**, *23*, 142–151. [CrossRef]
17. Islama, M.; RizwanAzhar, M.; Fredj, N.; Burleigh, T.D.; Oloyede, O.R.; Almajid, A.A.; Shah, S.I. Influence of SiO$_2$ nanoparticles on hardness and corrosion resistance of electroless Ni–P coatings. *Surf. Coat. Technol.* **2015**, *261*, 141–148. [CrossRef]
18. Gadharia, P.; Sahoo, P. Optimization of Coating Process Parameters to Improve microhardness of Ni–P–TiO$_2$ composite coatings. *Mater. Today Proc.* **2015**, *2*, 2367–2374. [CrossRef]
19. Chen, Y.; Hao, Y.; Huang, W. Corrosion behavior of Ni–P–nano–Al$_2$O$_3$ composite coating in the presence of anionic and cationic surfactants. *Surf. Coat. Technol.* **2017**, *310*, 122–128. [CrossRef]
20. Tamilarasan, T.; Rajendran, R.; Rajagopal, G.; Sudagar, J. Effect of surfactants on the coating properties and corrosion behaviour of Ni–P–nano–TiO$_2$ coatings. *Surf. Coat. Technol.* **2015**, *276*, 320–326. [CrossRef]

21. Afroukhteh, S.; Dehghaniann, C.; Emamy, M. Preparation of electroless Ni–P composite coatings containing nano-scattered alumina in presence of polymeric surfactant. *Prog. Nat. Sci. Mater. Int.* **2012**, *22*, 318–325. [CrossRef]

22. Nwosu, N.; Davidson, A.; Hindle, C.; Barker, M. On the Influence of surfactant incorporation during electroless nickel plating. *Ind. Eng. Chem. Res.* **2012**, *51*, 5635–5644. [CrossRef]

23. Sudagar, J.; Lian, J.; Sha, W. Electroless nickel, alloy, composite and nano coatings—A critical review. *J. Alloy. Compd.* **2013**, *571*, 183–204. [CrossRef]

24. Agarwala, R.C.; Agarwala, V. Electroless alloy/composite coatings: A review. *Sadhana* **2003**, *28*, 475–493. [CrossRef]

25. Sudagar, J.; Lian, J.S.; Jiang, Q.; Jiang, Z.H.; Li, G.Y.; Elansezhian, R. The performance of surfactant on the surface characteristics of electroless nickel coating on magnesium alloy. *Prog. Org. Coat.* **2012**, *74*, 788–793. [CrossRef]

26. Zielinska, K.; Stankiewicz, A.; Szczygie, I. Electroless deposition of Ni–P–nano–ZrO_2 composite coatings in the presence of various types of surfactants. *J. Colloid Interface Sci.* **2012**, *377*, 362–367. [CrossRef] [PubMed]

27. Der Ger, M.; Hwang, B.J. Effect of surfactants on codeposition of PTFE particles with electroless Ni–P coating. *Mater. Chem. Phys.* **2002**, *76*, 38–45. [CrossRef]

28. Mafi, I.R.; Dehghanian, C. Comparison of the coating properties and corrosion rates in electroless Ni–P–PTFE composites prepared by different types of surfactants. *Appl. Surf. Sci.* **2011**, *257*, 8653–8658. [CrossRef]

29. Chen, B.H.; Hong, L.; Ma, Y.; Ko, T.M. Effects of Surfactants in an Electroless Nickel–Plating Bath on the Properties of Ni–P Alloy Deposits. *Ind. Eng. Chem. Res.* **2002**, *41*, 2668–2678. [CrossRef]

30. Amell, A.; Muller, C.; Sarret, M. Influence of fluoro surfactants on the codeposition of ceramic nanoparticles and the morphology of electroless Ni–P coatings. *Surf. Coat. Technol.* **2010**, *205*, 356–362. [CrossRef]

31. Nwosu, N.O.; Davidson, A.M.; Hindle, C.S. Effect of Sodium Dodecyl Sulphate on the Composition of Electroless Nickel-Yttria Stabilized Zirconia Coatings. *Adv. Chem. Eng. Sci.* **2011**, *1*, 118–124. [CrossRef]

32. Mai, Y.; Zhou, M.; Ling, H.; Chen, F.; Lian, W.; Jie, X. Surfactant-free electrodeposition of reduced graphene oxide/copper composite coatings with enhanced wear resistance. *Appl. Surf. Sci.* **2018**, *433*, 232–239. [CrossRef]

33. Abdoli, M.; Sabour Rouhaghdam, A. Preparation and characterization of Ni–P/nanodiamond coatings: Effects of surfactants. *Diam. Relat. Mater.* **2013**, *31*, 30–37. [CrossRef]

34. Zarebidaki, A.; Allahkaram, S.R. Effect of surfactant on the fabrication and characterization of Ni–P–CNT composite coatings. *J. Alloy. Compd.* **2011**, *509*, 1836–1840. [CrossRef]

35. Bulasara, V.K.; Babu, C.S.N.M.; Uppaluri, R. Effect of surfactants on performance of electroless plating baths for nickel-ceramic composite membrane fabrication. *Surf. Eng.* **2012**, *28*, 44–48. [CrossRef]

36. Tamilarasan, T.R.; Rajendran, R.; Sivashankar, M.; Sanjith, U.; Rajagopal, G.; Sudagar, J. Wear and scratch behaviour of electroless Ni-P-nano-TiO_2: Effect of surfactants. *Wear* **2016**, *346*, 148–157. [CrossRef]

37. Song, X.; Jiang, N.; Li, Y.; Xu, D.; Qiu, G. Synthesis of CeO_2-coated SiO_2 nanoparticle and dispersion stability of its suspension. *Mater. Chem. Phys.* **2008**, 128–135. [CrossRef]

38. Zuo, D.; Tian, G.; Li, X.; Chen, D.; Shu, K. Recent progress in surface coating of cathode materials for lithium ion secondary batteries. *J. Alloy. Compd.* **2017**, *706*, 24–40. [CrossRef]

39. Zuo, D.; Tian, G.; Chen, D.; Shen, H.; Lv, C.; Shu, K.; Zhou, Y. Comparative study of Al_2O_3-coated $LiCoO_2$ electrode derived from different Al precursors uniformity, microstructure and electrochemical properties. *Electrochim. Acta* **2015**, *178*, 447–457. [CrossRef]

40. Liu, P.; Zhu, Y.; Zhang, S. Hydrophilicity characterization of Al_2O_3-coated MoS_2 particles by using thin layer wicking and sessile drop method. *Powder Technol.* **2015**, *281*, 83–90. [CrossRef]

41. Liu, Y.; Zhu, Y.; Liu, P.; Liu, T. Surface coating and application in plating of MoS_2 powders with Al_2O_3. *China Surf. Eng.* **2012**, *25*, 97–102. (In Chinese)

42. Liu, T.; Zhu, Y.; Liu, Y.; Zhang, S. Preparation and properties of Ni–P–MoS_2/Al_2O_3 composite coating. *Lubr. Eng.* **2013**, *38*, 46–50. (In Chinese)

43. Liu, Y.F.; Zhu, Y.W.; Liu, T.T.; Zhang, S.W.; Peng, Y. Friction and wear properties of Ni–P electroless composite coatings with core-shell nanodiamond. *Tribology* **2013**, *33*, 267–274.

44. Gan, B.K.; Madsen, I.C.; Hockridge, J.G. In situ X-ray diffraction of the transformation of gibbsite to a-alumina through calcination: Effect of particle size and heating rate. *J. Appl. Crystallogr.* **2009**, *42*, 697–705. [CrossRef]

45. Bigdeli, F.; Allahkaram, S.R. An investigation on corrosion resistance of asapplied and heat treated Ni–P/nanoSiC coatings. *Mater. Des.* **2009**, *30*, 4450–4453. [CrossRef]
46. Apachitei, I.; Tichelaar, F.D.; Duszczyk, J.; Katgerman, L. The effect of heat treatment on the structure and abrasive wear resistance of autocatalytic NiP and NiP–SiC coatings. *Surf. Coat. Technol.* **2002**, *149*, 263–278. [CrossRef]
47. Hu, X.G.; Cai, W.J.; Xu, Y.F.; Wan, J.C.; Sun, X.J. Electroless Ni–P–(nano-MoS$_2$) composite coatings and their corrosion properties. *Surf. Eng.* **2009**, *25*, 361–366. [CrossRef]
48. Elansezhian, R.; Ramamoorthy, B.; Nair, P.K. The influence of SDS and CTAB surfactants on the surface morphology and surface topography of electroless Ni–P deposits. *J. Mater. Proc. Technol.* **2009**, *209*, 233–240. [CrossRef]
49. Moonir-Vaghefi, S.M.; Saatchi, A.; Hejazi, J. Deposition and properties of electroless nickel–phosphorus–molybdenum disulfide composites. *Met. Finish.* **1997**, *95*, 46, 48, 50–52. [CrossRef]

coatings

MDPI

Article

Structure-Property Relationships in Suspension HVOF Nano-TiO$_2$ Coatings

Feifei Zhang [1,2], Shuncai Wang [1,*], Ben W. Robinson [2], Heidi L. de Villiers Lovelock [3] and Robert J.K. Wood [1]

[1] National Centre for Advanced Tribology at Southampton (nCATS), School of Engineering, University of Southampton, Southampton SO17 1BJ, UK
[2] Surface Engineering, TWI Limited, Cambridge CB21 6AL, UK
[3] Oerlikon Metco WOKA GmbH, 36456 Barchfeld-Immelborn, Germany
* Correspondence: wangs@soton.ac.uk; Tel.: +44-238-059-4638

Received: 12 July 2019; Accepted: 7 August 2019; Published: 9 August 2019

Abstract: Hardness and tribological properties of microstructured coatings developed by conventional thermal spraying are significantly affected by the feedstock melting condition, however, their effect on the performance of nanostructured coatings by suspension high velocity oxy-fuel (HVOF) are inconclusive. In this work, nano-TiO$_2$ coatings with different degrees of melting (12%, 51%, 81%) of nanosized feedstock were deposited via suspension HVOF spraying, using suspensions with a solid content of 5 wt.%. All the coatings produced had dense structures without visible pores and cracks. Two TiO$_2$ crystal structures were identified in which the rutile content of the coatings increased with increased feedstock melting. Their mechanical, friction and wear behaviours largely relied on the extent of melting of the feedstock. The coating composed of mostly agglomerate particles (12% melted particles) had the lowest coefficient of friction and wear rate due to the formation of a smooth tribo-film on the wearing surface, while the coating composed of mostly fully melted splats (81% melted particles) presented the highest coefficient of friction and low wear rate, whose wear mechanism was dominated by abrasive wear and accompanied by the formation of cracks.

Keywords: HVOF; suspension; TiO$_2$; thermal spray; friction and wear behaviour

1. Introduction

In recent years, a modified thermal spraying process using a fine suspension of submicron or even nanostructured powders in a liquid phase as the feedstock material has gained an increased level of interest in the scientific world [1]. The use of suspensions opens up an entirely new class of spray materials for the production of nanostructured coatings by thermal spray technologies, as conventionally only tens of micron-sized powders can be used with a standard powder feeding device [2]. The introduction of the liquid phase in the suspension provides good flowability and therefore allows direct feeding of nanosized or submicron-sized feedstock, which enables the fabrication of finely structured coatings [1–3]. Suspension thermal sprayed coatings exhibit refined splats whose size is at least one order of magnitude smaller than that of the conventional thermal sprayed coatings [4], and the coating thickness can be controlled in a range from a few µm to several mm [1].

Thermal sprayed coatings are widely used in industry to improve the tribological properties in a number of applications, such as rolls, pump bodies and plungers, as well as machinery parts [3,5]. Recent studies have shown that nanostructured coatings exhibit outstanding properties, such as better sliding wear resistance than those of conventional ones [3,5–9]. For example, the wear rate of nanostructured Y$_2$O$_3$-ZrO$_2$ (YSZ) coatings, which had higher hardness, lay between 25% and 40% of that of conventional coatings [6,9]. It has also been observed that nanostructured coatings exhibited enhanced crack propagation resistance against wear, even when they were not harder than the corresponding conventional coatings. For example, an improvement of three to four times of wear resistance under dry sliding conditions was observed for nanostructured Al_2O_3-13 wt.% TiO_2 coatings when compared with optimized microstructure coatings, even though they have a lower hardness [7,8]. Therefore, the overall hardness and wear resistance for nanostructured thermal sprayed coatings are not always correlated.

Generally, the hardness of materials is the most critical factor on wear resistance, although other factors, including ductility, toughness and microstructure, also play a role in the wear process [7,10]. However, there is still no established correlation between the amount of nanostructured zones embedded in the microstructure and the coating performance [3]. The tribological behaviour of the nanostructured coatings are more related to the amount of unmelted powder incorporated into the final coating [11]. The study on crack growth resistance of nanostructured thermal sprayed Al_2O_3-TiO_2 coatings showed that 60% or more of the crack arrest events are trapped within the partially-melted regions and deflected at the interface between partially-melted and fully-melted regions, compared to only 3%–12% in fully-melted splats [12]. To the extent of our knowledge, no detailed studies regarding the influence of melting conditions on the friction and wear behaviour of thermal sprayed nanostructured coatings have been published in the literature.

In the present study, three kinds of nanostructured TiO_2 coatings with different melting conditions of feedstocks were fabricated by varying the deposition parameters of the high velocity oxy-fuel (HVOF) process in order to ascertain how different coating structures affect the coating tribological performance. The constituent phases, microstructure, mechanical properties, friction and wear behaviour under dry sliding contact conditions of the coatings were examined in detail using a number of characterization techniques.

2. Materials and Methods

2.1. Suspension Preparation

A commercial TiO_2 nanopowder (Aeroxide P25, Degussa-Evonik, Hanau, Germany) was used as the TiO_2 feedstock. The TiO_2 suspensions were prepared in-house and consisted of 5 wt.% solid powder and 95 wt.% solvents. The solvents were mixtures of H_2O and isopropanol (Table 1) and acted as a carrier during feedstock feeding.

Table 1. Deposition parameters of suspension high velocity oxy-fuel (HVOF) sprayed TiO_2 coatings.

Label	Suspension Feed Rate, mL/min	Solvent, *v/v*	Spray Distance, mm	Fuel
S1	20	H_2O:isopropanol = 9:1	130	Propylene
S2	20	H_2O:isopropanol = 10:0	100	Hydrogen
S3	20	H_2O:isopropanol = 9:1	150	Hydrogen

2.2. Spray Process

The as-prepared suspensions were deposited onto a commercially available stainless steel (AISI Grade 304) substrate (25 mm × 25 mm × 1.5 mm). All the samples were grit blasted with 100 mesh fused alumina abrasive prior to HVOF spraying. The coating deposition process was carried out at TWI Limited in Cambridge, UK, using a UTP Top Gun torch with a 22 mm long combustion chamber and a 135 mm long expansion nozzle mounted on an OTC AII-V20 robot. The suspension feed rate was controlled using an ISCO 260D syringe pump. The suspension was injected perpendicularly into the flame using a 0.3 mm nozzle mounted on the top of the torch at the combustion exit (Figure 1). Since the feedstock particle size used in suspension sprayed was much smaller compared with conventional spraying, there was much lower feedstock momentum and thermal inertia. The spray distance was reduced to compensate for the decreased particle kinetic energy. Thus, the heat flux from the flame into the substrate was much higher, at least one order of magnitude more than the typical value for conventional spraying at similar conventional powder feed rates [13]. A water-cooling system was attached to the back of the substrates to extract heat and prevent distortion of the substrates by keeping the substrate temperature at a constant 55 °C. Compressed air was applied to further cool down the samples after spraying. A schematic of the suspension HVOF system is shown in Figure 1. The main suspension HVOF spraying parameters are listed in Table 2 [14,15].

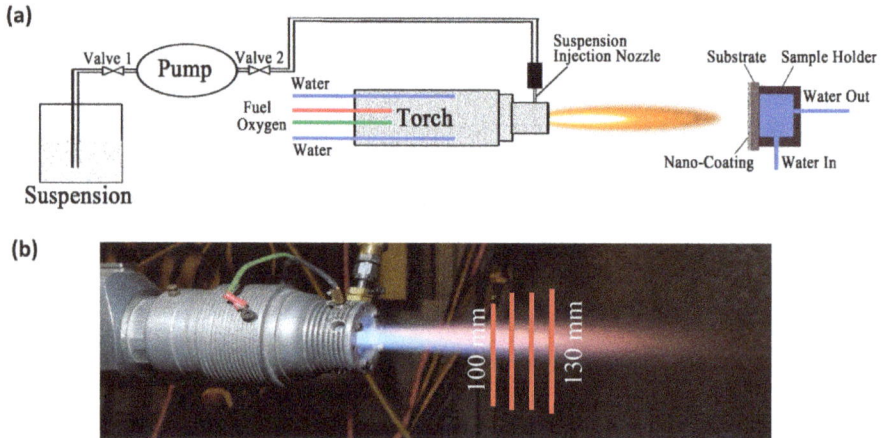

(a)

(b)

Figure 1. (a) Schematic of the suspension HVOF system for nanostructured TiO$_2$ coating. (b) Digital image of the HVOF propylene flame with suspension.

Table 2. Main suspension HVOF spraying parameters.

Parameters	Value
Pass spacing, mm	2
Torch linear velocity, mm/s	600
Torch cooling system	Water cooling
Combustion chamber length, mm	135
Profile of suspension nozzle, mm	0.3/orifice
Number of passes	15
Flame condition 1	–
Propylene flow rate, slpm	80.5
Oxygen flow rate, slpm	280.0
Flame condition 2	–
Hydrogen flow rate, slpm	788.0
Oxygen flow rate, slpm	264.0

2.3. Coating Characterisation

The morphology of the samples (feedstock, surface, cross-section and worn scar of the coating) was characterised using scanning electron microscopy (SEM) (JEOL JSM 6500F, Tokyo, Japan). Cross-sectional samples were cold-mounted in resin, ground with SiC grit papers in stages from 120 down to 4000 mesh and finally polished with 0.4 μm Al$_2$O$_3$ slurry. All SEM images in the study were secondary electron images. A non-contact 3D optical profilometry (Alicona InfiniteFocus SL, Raaba/Graz, Austria) was used to characterise the surface roughness of the as-deposited specimens (20×). At least three readings were taken for each sample and the average value was recorded.

The crystalline phases of the coatings were determined by X-ray diffraction (XRD) using a Bruker D2 PHASER diffractometer (Billerica, MA, USA) in the reflection mode with Cu-Kα radiation (λ = 0.154 nm). The scan step was 0.02°, with a step time of 0.5 s in the 20–80° 2θ range. Peaks of phases were analysed by Jade XRD software (version 5.0). Titanium oxide has two structures—anatase and rutile. Both anatase and rutile are tetragonal structures with different *c/a* ratios of 2.52 and 0.64, respectively, and different densities of 3.915 and 4.276 g/cm^3, respectively. Anatase has a relative larger spacing which may accompany with easy slide. The volume percentage of rutile (C_R) was determined according to the following equation [16]:

$$C_R = \frac{13\,I_R}{8I_A + 13\,I_R} \tag{1}$$

where I_A and I_R are the X-ray intensities of the anatase (101) and the rutile (110) peaks, respectively.

Nano-indentation testing was performed by NanoTest Vantage (Micro Materials Ltd., Wrexham, UK) onto a polished cross-sectional surface of suspension HVOF TiO$_2$ coating using a three-sided pyramidal Berkovich diamond indenter tip with a diameter of 200 nm. A fixed penetration depth of 300 nm, loading rate of 0.4 mN/s and holding time of 40 s were used for testing. Multiple indentations separated by 6 μm were performed for each load value. The hardness and the elastic modulus were recorded by the nano-indentation software, and the Poisson's ratio was assumed to be 0.27 [17]. Five readings were taken for the final results.

Reciprocating wear testing was carried out on the surface of as-deposited coatings TE77 (Phoenix, UK) to determine the friction and wear behaviour of solid surfaces in sliding contact. All the tests were performed under dry sliding conditions under a constant load of 5 N (the initial Hertzian contact pressure was 1.27 GPa), at an ambient temperature of 23 ± 1 °C and 60 ± 1% relative humidity (ASTM G133–05) [18]. The stroke length was 2.69 mm with a sliding frequency of 1 Hz (the average sliding speed was 5.38 mm/s), and the total sliding distance was 3.228 m. A sintered Al$_2$O$_3$ ball (manufacturer's nominal hardness of 19 GPa) with diameter of 6 mm was used as the counter body. A piezo electric transducer was used to measure the friction force. The coefficient of friction and the sliding time were recorded automatically during the test. At the end of the test, the sample was cleaned by compressed air flow to remove loose debris. The track profile was acquired by a non-contacting 3D microscope (Alicona InfiniteFocus SL, Raaba/Graz, Austria), and at least five profile measurements were taken for each wear track. The corresponding specific wear rate was calculated from the equation:

$$K = \frac{V}{SF} \tag{2}$$

where V is the wear volume in mm^3, S is the total sliding distance in metres and F is the normal load in newton.

3. Results

3.1. Microstructure

The prepared suspension was dried and the particles were observed by SEM (Figure 2a). The nanoparticles were distributed uniformly within the suspension, with small agglomerates only about several hundreds of nanometres in size (circled in Figure 2a). No stratification could be observed in the sedimentation test even after the TiO$_2$ suspension had been left untouched for 36 h (inset in Figure 2a). In order to further characterise the stability of the 5 wt.% TiO$_2$ suspension, the zeta potential was monitored over a period of 120 min (Figure 2b). When the zeta potential is higher than 30 mV or lower than −30 mV, it is accepted that the suspension can resist strong agglomeration and can be electrically stabilized [19]. During the first 60 min, the electrokinetic potential varied between −41 and −57 mV, and then tended to stay stable with little variation between −49 and −56 mV. This was because the electrical charge at the double layer of particles in freshly prepared suspensions is not electrically stabilized [20] and the equilibrium state between particles and dispersed solution can be reached in 1 h for suspensions with a solid content of 5 wt.%.

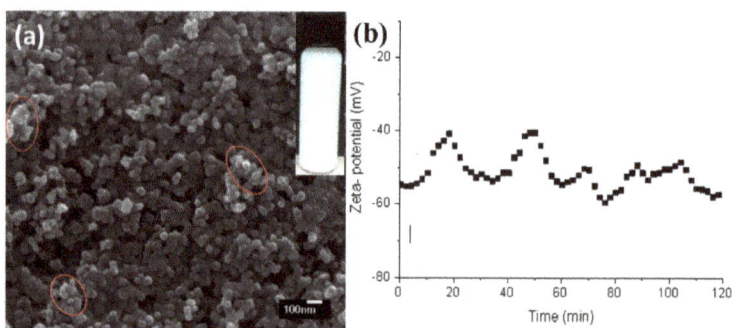

Figure 2. (**a**) Scanning electron microscopy (SEM) image of TiO$_2$ nanoparticles in the suspension (the inset is the digital image of the sedimentation test of suspension after 7 days); (**b**) zeta potential changes with time.

Figure 3 shows the surface and cross-section microstructures of suspension HVOF TiO$_2$ coatings deposited using different parameters. Typical surface features include re-solidified particles, semi-melted, fully-melted and/or agglomerated particles caused by different thermal histories of each individual particle/agglomerate which then lead to different impacting behaviours onto the substrate [14]. The as-deposited coatings generally have very low surface roughness and exhibit smooth surface features compared with that of conventional thermal sprayed coatings [21]. The average surface roughness (R_a) for the three coatings (as shown in Table 3) ranged from 0.53 to 1.18 µm. The cross-sections of suspension HVOF coatings clearly show that they possessed "bimodal" distributed microstructures, where fully-melted zones (bright zones) correspond to the rutile phase and agglomerate zones (grey zones) correspond to the anatase phase, based on Raman spectra analysis as reported by in [22]. Furthermore, the suspension HVOF TiO$_2$ coatings exhibited a dense structure, with no visible pores and no obvious defects when observed by SEM. The coatings adhered well to the substrate along the surface profile. None of the coatings appeared to have the typical micro-cracks frequently observed in conventional thermal sprayed coatings, with these being the result of the relaxation of thermal stresses generated during processing associated with significant convective heat input [23]. The coating thicknesses were 5.6 ± 1.7, 15.5 ± 2.1 and 8.1 ± 2.8 µm for coatings S1, S2 and S3, respectively.

The cross-sectional images were converted from greyscale images into binary images using a threshold to evaluate their extents of melting, as shown in Figure 4. The extent of feedstock melting was ca. 12%, 51% and 81% for coatings S1, S2 and S3, respectively. The crystalline structure of the coating was analysed by X-ray diffraction patterns, as shown in Figure 5. The P25 nanopowder was composed of anatase and rutile phases in the proportion 81:19. No other phases were observed in the coatings compared with P25 feedstock, except the peaks at around 44.5° that revealed the austenite phase of the stainless steel substrate due to the low thickness of the coating. An increase of rutile peak intensity was observed in the XRD patterns for the coatings compared with P25 powder. This indicated that anatase-to-rutile phase transformation occurred during the deposition process because of heat transfer from the HVOF flame jet. The rutile content (C_R) was 23.6%, 63.0% and 70.0% for coatings S1, S2 and S3, respectively, a tendency consistent with the extent of feedstock melting observed by SEM. More fully-melted regions in the coating led to higher C_R, which is consistent with results in [22].

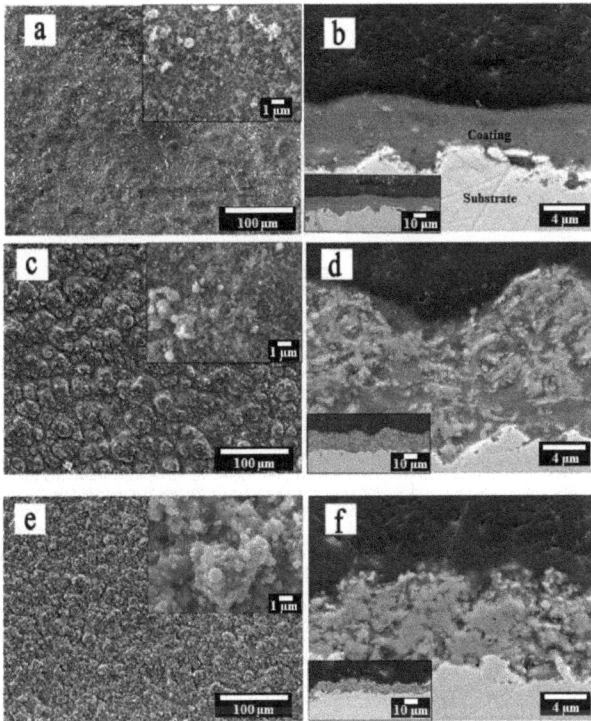

Figure 3. SEM images of surface and cross-sectional coating morphologies: (**a**,**b**) S1, (**c**,**d**) S2 and (**e**,**f**) S3.

Table 3. Values of the coefficient of friction and the specific wear rate of different samples.

Samples	Hardness H_v, GPa	Surface Roughness R_a, µm	Wear Rate, $\times 10^{-7}$ mm^3/Nm	Coefficient of Friction
304SS	4.7 ± 0.3	0.56 ± 0.14	5.13 ± 0.04	0.55 ± 0.05
S1	2.1 ± 0.3	0.53 ± 0.14	0.83 ± 0.03	0.35 ± 0.02
S2	4.0 ± 0.9	1.18 ± 0.18	5.13 ± 0.13	0.48 ± 0.04
S3	7.8 ± 0.4	0.96 ± 0.17	1.77 ± 0.05	0.62 ± 0.03

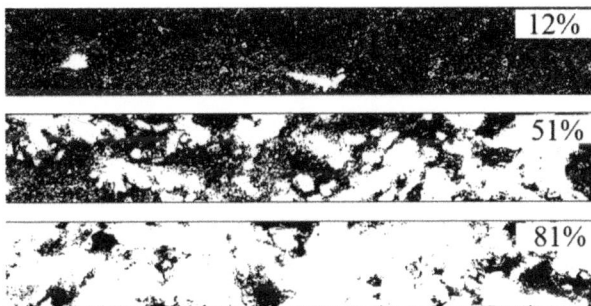

Figure 4. Binary images of suspension HVOF TiO_2 coatings converted from their greyscale images by image thresholding in order to assess the melting percentages: S1 (12%), S2 (51%) and S3 (81%).

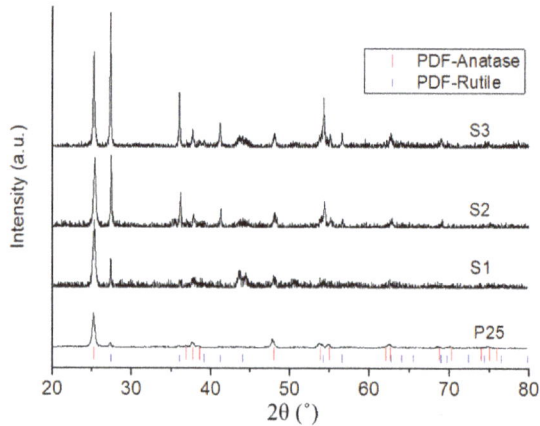

Figure 5. X-ray diffraction (XRD) analysis of suspension HVOF TiO$_2$ coatings.

3.2. Mechanical Properties

The hardness and elastic modulus in different melting zones of the coating cross-section were evaluated by nano-indentation with a depth of 300 nm. Hardness is an indicator of the irreversible or plastic deformation behaviour of the test material [24]. The average hardness values for the agglomerated zone (mostly the anatase phase) and fully-melted zone (mostly rutile) were 2.1 and 7.8 GPa, respectively (Figure 6). These values are lower than those reported in the literature for suspension thermal sprayed TiO$_2$ coatings, which were above 8.5 GPa [25,26]. This can be ascribed to the lower organic content of the suspension and the lower suspension feed rate used in this study, with both leading to less melting of the feedstock. For example, some of the TiO$_2$ splats had a fully-melted solid core with a partially-melted or agglomerated interior structure (as circled in Figure 7) and the loosely accumulated splats tended to build more pores. Hardness measurements have been found to be sensitive to these built flaws generated in the coating, particularly porosity [27]. The elastic modulus of various zones for suspension HVOF TiO$_2$ coating lay between 27% and 50% of that of the bulk TiO$_2$ (282.0 GPa), and the elastic modulus of the fully-melted zone (135 GPa) was only slightly lower than that of a coating produced using a conventional HVOF process (164 GPa) [26]. This was within the expected range (20%–50%) for thermal sprayed coatings when compared with bulk material [28,29].

The variations of both hardness (*H*) and elastic modulus (*E*) across different melted zones show that they are closely related to the coating microstructure. Both of them increased as the amount of fully-melted particles increased (Figure 6). The fully-melted zone had the highest hardness and elastic modulus values due to its lower porosity. The lower values of elastic modulus and hardness compared to that of equivalent bulk materials can be attributed to the unique microstructure of suspension thermal sprayed deposits, which is typically composed of fully-melted, partially-melted and agglomerated particles.

The extent of the recovery during nano-indentation depends on the hardness-to-elastic modulus ratio (*H/E*). The *H/E* is an indicator of a material's capacity to absorb or dissipate energy, which is lower for more plastic materials and becomes higher for more elastic materials. The fully melted zones had the highest elasticity index and showed superior elastic property, while the agglomerated zones were more plastic (e.g., Figures 6 and 7). SEM examination of indents on a number of melted zones revealed that there were no cracks present around the indents (Figure 7). All the indentation areas were smooth, without any sign of incipient crack formation or material accumulation. A shallow and small indentation impression was observed for fully melted zones, compared with much larger ones for agglomerated zones (e.g., Figure 7a,c).

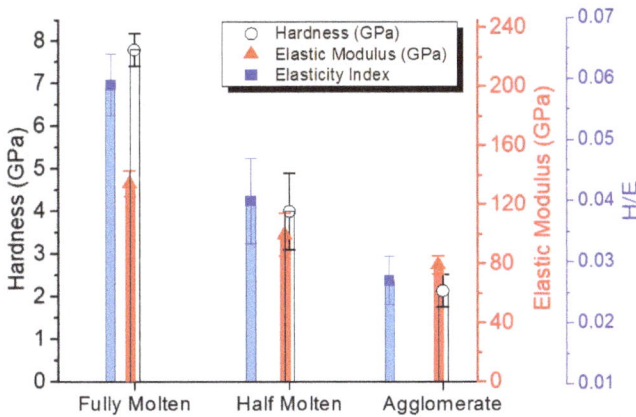

Figure 6. Comparison of nano-indentation results (*H*, *E* and *H/E*) on different zones of the suspension HVOF TiO$_2$ coating (S2).

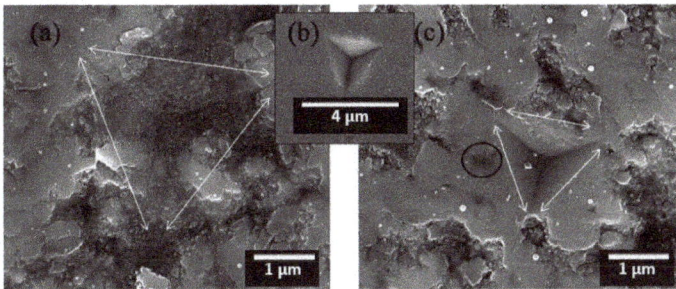

Figure 7. Morphologies of nano-indents: (**a**) region of agglomerated particles, (**b**) substrate-304 SS and (**c**) region of fully melted splats.

3.3. Friction and Wear Behaviour

As the thickness of the deposited coatings was very thin (less than 20 µm, compared with several hundreds of micrometres for conventional thermal sprayed coatings), the reciprocating wear test only ran for 600 s to avoid the effect of the substrate. The as-sprayed surface roughness did not seem to be critical for wear and friction performance of the developed coatings in the study when their R_a varied between 0.53 and 1.18 µm (Table 3). Suspension HVOF TiO$_2$ coatings prepared in this study had various wear rates, in the range of 0.83–5.13 × 10^{-7} mm^3/Nm. The coefficient of friction varied from 0.35 to 0.62, which is lower than that of conventional plasma sprayed and suspension HVOF TiO$_2$ coatings obtained from dry sliding ball-on-disc tests using the same counterpart 6 mm Al$_2$O$_3$ ball but with a load of 2 N as found in the literature, i.e., ca. 0.90 [30]. With the increase of the melting extent, the friction coefficients increased though the wear resistance varied differently. The coefficient of friction curves of different samples (Figure 8) indicated that all the suspension HVOF TiO$_2$ coatings had smoother curves compared with that of 304SS, regardless of the extent of the melting of the feedstock. However, the coatings with different melting conditions of feedstock had very different coefficient of friction values. The coating with mostly agglomerated particles (S1) showed the lowest coefficient of friction, and the coating with mostly fully melted particles (S3) presented the highest coefficient of friction. It can be assumed that coefficient of friction is related with particle melting extent, which is then reflected by phase structure, i.e., anatase has a lower coefficient of friction than rutile.

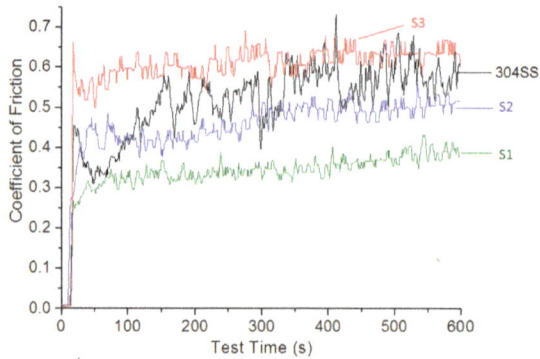

Figure 8. Evolution of the coefficient of friction curves of suspension HVOF TiO$_2$ coatings.

In order to further study the wear mechanism, the surface morphologies of the wear scars were examined by SEM. Uncoated 304SS substrate had a specific wear rate of 5.13 ± 0.04 × 10^{-7} mm^3/Nm (Table 3). Its coefficient of friction varied between 0.38 and 0.72 (Figure 8), consistent with the typical values for the contact between ceramic and metallic materials [31]. The large fluctuations in the data were typical for poor tribological materials with significant adhesive wear and stick-slip tendencies. Wear scar analysis confirmed that a large amount of wear debris from the substrate adhered to the contact surfaces of both the substrate and the counterpart Al$_2$O$_3$ ball (Figure 9). Severe plastic deformation of steel wear debris and the generation of a series of grooves on the wear scar surface indicated that two-body ploughing also occurred during the wear process (Figure 9a,b). The presence of a high amount of oxygen (overlap with Cr peak) on the 304SS surface (compared with EDS analysis of bare 304SS substrate in Figure 9d) indicated strong oxidation due to elevated temperatures between the contact surfaces (Figure 9c). The wear process of uncoated 304SS substrate was dominated by adhesive wear and two-body ploughing and had a typically fluctuating coefficient of friction curve.

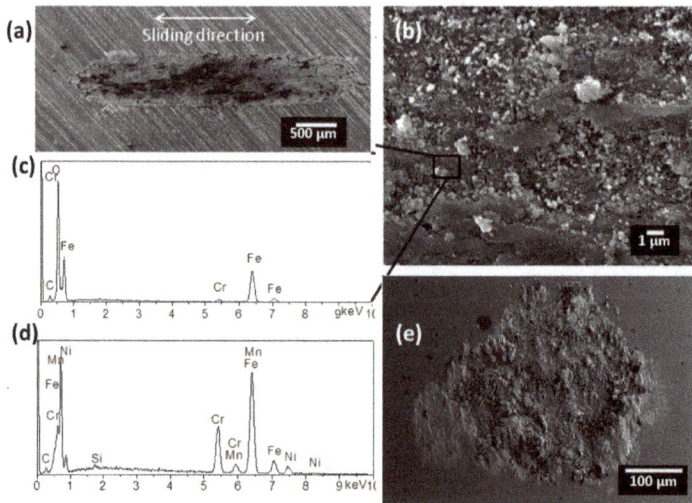

Figure 9. Wear track of 304SS under (**a**) low magnification and (**b**) high magnification; (**c**) EDS analysis of area in image (**b**); (**d**) area EDS analysis of as-received bare 304SS substrate; (**e**) wear scar of counterpart Al$_2$O$_3$ ball.

It is widely considered within the thermal spray community that a higher amount of fully melted splats is essential for improving the strength of thermal spray coatings [32] and, as a result, coatings composed of mostly agglomerate particles have received much less attention, especially for tribological applications. In this study, coating S1, with a feedstock melting percentage of 12% (C_R = 23.6%), showed a very low coefficient of friction and low wear rate (Figures 8 and 10). High magnification examination of the coating in the cross-section showed that it was mainly composed of agglomerated spherical granular TiO_2 particles (Figure 10a). From the SEM analysis, the wear scar was small and had a smooth surface without wear debris (Figure 10b,c). Only a few small localized cracks shorter than 3 μm (as observed under SEM) and perpendicular to the sliding direction could be seen on the worn surface (Figure 10c). It is assumed that the existence of agglomerated granular particles plays an important role in arresting and deflecting crack propagation during wear. The act of applying the test load led to a densification of the coating on the worn surface and a layer of compact tribo-film was therefore formed by the deformation of splats, which was mainly TiO_2 (Figure 10a,c), which provided a low coefficient of friction. The EDS analysis revealed the exposure of substrate material (iron) as the coating was worn through (Figure 10d), indicating that the tribo-film can be continuously formed and worn off because of the much higher hardness of the Al_2O_3 ball compared with that of the coating. Only very little TiO_2 debris was found adhering onto the surface of the Al_2O_3 ball because of the formation of tribo-film (Figure 10e). Therefore, the improved friction and wear properties of the coating composed of mostly agglomerated feedstock can be ascribed to the effective hindrance of crack generation and propagation of the granular feedstock, which leads to enhanced integral deformation resistance of the nanostructured coatings.

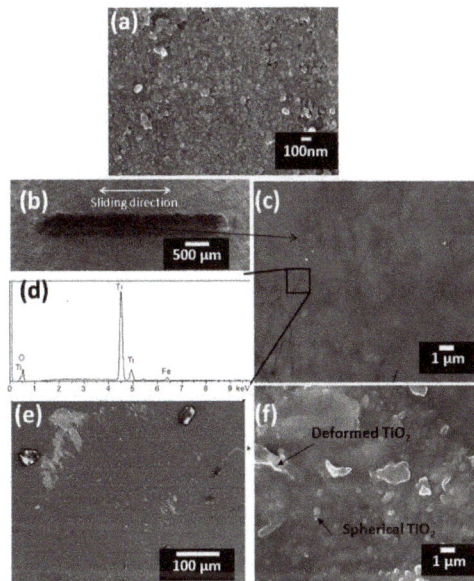

Figure 10. Coating S1: (**a**) cross-sectional SEM view of coating S1, wear track under (**b**) low magnification and (**c**) high magnification; (**d**) EDS analysis of area in image; wear scar of counterpart Al_2O_3 pin under (**e**) low magnification and (**f**) high magnification.

With increased extents of feedstock melting, the coating S2 had a higher coefficient of friction and complex wear behaviour (Figures 8 and 11). Unlike the previous coating (S1) composed of mostly agglomerated particles, a considerable amount of pulled-out debris was found on the worn surface of S2, especially along the edges of the wear scar (Figure 11a). From the enlarged image of the wear scar, the generated cracks were much bigger compared with that of coating S1, which was due to the increased level of the extent of melting and the corresponding decrease in plasticity (Figure 11b). The inhibition and deflection effects of preserved nanostructured zones on the crack propagation were clearly observable, confirming that the existence of nano-zones can act as crack arresters (i.e., energy absorbers) and improve the toughness of the coating. A similar phenomena was described in [33] by applying a micro-indentation test on a nanostructured HVOF TiO_2 coating. The EDS analysis showed that the worn surface of the coating was mostly TiO_2 with a small amount of iron from the substrate, whilst no aluminium, which would indicate counterpart alumina material transfer, could be detected (Figure 11c). Fine and irregular wear debris composed of sintered big agglomerates and fully-melted splats were observed on the contact surfaces (Figure 11d,e). Since a coating that is composed of half-melted feedstock will have an increased level of build defects such as weak intersplat boundaries, the vertically applied load can lead to the displacement of loosely attached TiO_2 particles on the coating surface asperities. These fragments can be transferred to the edges of the wear scar or remain at the interface and act as a third body. The trapped debris between the contact bodies accelerates particle exfoliation and the delamination of the tribo-film during the reciprocating sliding wear process. More wear debris could be formed as a resulted of surface fatigue, brittle fracture and crumbling. The specific wear rate of coatings with moderate extents of feedstock melting can therefore be very high when such coatings present high levels of defects.

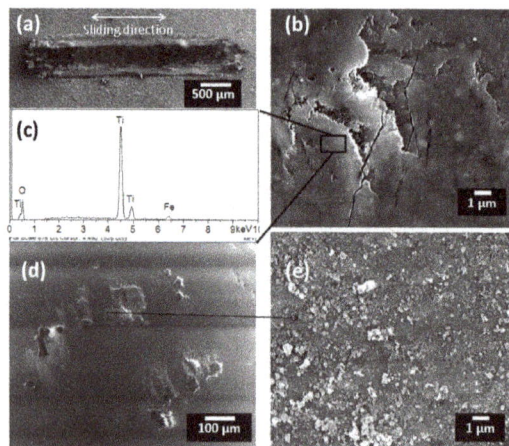

Figure 11. Wear track of coating S2 under (**a**) low magnification and (**b**) high magnification; (**c**) EDS analysis of area in image (**b**); wear scar of counterpart Al_2O_3 pin under (**d**) low magnification and (**e**) high magnification.

For coating S3 that was composed of mostly fully melted feedstock, its coefficient of friction was higher than the previous two types of coating (Figures 8 and 12). From Figure 12a, the wear scar was clean and no obvious wear debris had accumulated on the surface. It can be clearly seen from the enlarged image of the wear track surface that the fatigue crack propagation runs perpendicular to the sliding direction (Figure 12b), which was caused by surface tensile stresses. From Section 3.2, the overall hardness of the coating increased with the increase of the number of fully melted splats, and at the same time the plasticity decreased. Furthermore, the increase of the proportion of fully melted splats caused a loss of nano-features in the coating, which restrained the formation of the tribo-film. These led to a decrease in crack propagation resistance and a correspondingly high coefficient of friction. The presence of aluminium in the EDS spectrum on the worn surface indicates that material transfer occurred between the coating and the Al_2O_3 ball surface (Figure 12c). Further investigation of the surface of the counterpart Al_2O_3 ball shows that many detached fragments were adhered to the edge of the wear track and the ball surface had experienced significant abrasion in the centre of the wear scar (Figure 12d,e). In summary, the coating composed of mostly fully-melted feedstock still had good wear resistance due to its high hardness, albeit with a higher coefficient of friction (Table 3).

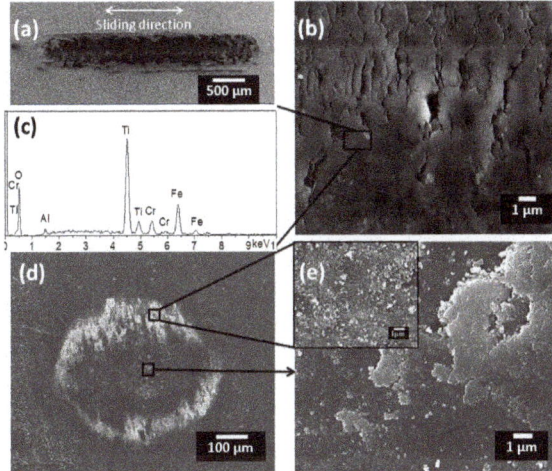

Figure 12. Wear track of coating S3 under (**a**) low magnification and (**b**) high magnification; (**c**) EDS analysis of area in image (**b**); wear scar of counterpart Al_2O_3 pin under (**d**) low magnification and (**e**) high magnification.

4. Discussion

From the information above, the friction and wear responses of suspension HVOF TiO_2 coatings are difficult to predict by adopting the principles of wear behaviour that are often used for conventional engineering materials. An increase in hardness is often associated with increased resistance in sliding, low-angle erosive or abrasive wear conditions for conventional materials within a particular material category [2,34], but it does not always improve the wear resistance of thermal sprayed coatings because of the complexity of their structure, which contains different microstructural features, including pores, cracks and fragments [5,35]. The existence of cracks and horizontal pores are ideal "shear faults" for plastic deformation and are favourable crack initiation points [36]. During the wear process, the vertically applied load can cause displacement of deposited splats that are mechanically adhered to each other. This debris trapped between the contact surfaces causes three-body abrasion and thus aggravates the wear of the coating. Nano-indentation demonstrated that the properties (hardness, elastic modulus, and elastic index) for the suspension HVOF TiO_2 coating were dissimilar for zones with different extents of melting of the feedstock (Figure 6). The more melted microstructure contained more rutile and had higher hardness, while the surrounded agglomerated zone consisted mostly of anatase and was softer [25]. Higher hardness and stiffness leads to a more brittle nature of the coatings that have more melting [37], whilst the higher induced shear force during wear causes a higher coefficient of friction. From wear tests, when the coating had a negligible level of built defects, the fully melted splats improved the cohesion strength and hardness, which thus led to good wear resistance, and the agglomerated zone was beneficial in getting a low coefficient of friction due to its good plasticity. In general, the wear mechanism of suspension HVOF TiO_2 coatings always involved the formation of a smooth and dense surface film (tribo-film) on the surface of the wear track and its progressive delamination and removal caused by wear debris and surface fatigue during wear (Figure 11). The observed results for suspension HVOF TiO_2 coatings in the study prove that the increase in friction appears to be related to the extent of melting of feedstock, i.e., increased levels of the rutile phase.

A further experiment has been carried out to confirm the relationship between coefficient of friction and rutile content. A series of coatings were deposited using different spray distances with propylene as fuel and a suspension feed rate of 20 mL/min. When increasing spray distances from 100 mm, the impacting speed of nanoparticles onto the substrate and flame temperature would become lower (Figure 1b). Coating microstructures, as shown in Figure 13, indicates different levels of feedstock melting conditions under each spray distance, and the extent of melting decreases when spray distance increases. The C_R, wear rates and friction coefficients are listed in Table 4. These results provide evidence that the agglomerated zone (mainly anatase) is beneficial for a low coefficient of friction and improved wear resistance. This could be explained by the easy separation of the anatase tribofilm along its spacious c-axis. However, further work is needed to verify this.

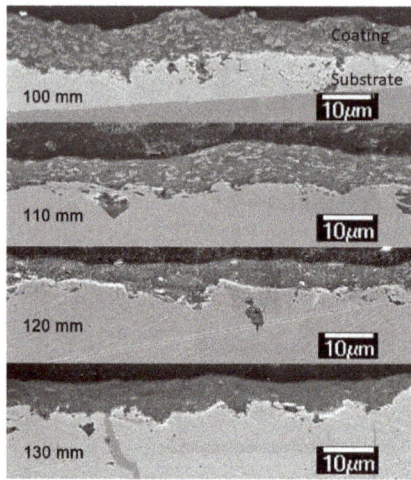

Figure 13. Influence of spray distance on the coating microstructure.

Table 4. List of properties of coatings with different C_R (H_2O:isopropanol = 9:1).

Spray Distance	C_R, %	Surface Roughness, μm	Coefficient of Friction	Specific Wear Rate, $\times 10^{-7}$ mm³/(N m)
100 mm	58	0.72 ± 0.10	0.68 ± 0.04	2.47 ± 0.07
110 mm	53	0.44 ± 0.15	0.54 ± 0.05	2.01 ± 0.09
120 mm	45	0.67 ± 0.08	0.38 ± 0.03	0.64 ± 0.07
130 mm	41	0.53 ± 0.14	0.36 ± 0.02	0.83 ± 0.02

5. Conclusions

Nanostructured TiO_2 coatings were deposited onto a stainless steel substrate (304SS) by controlling the level of heat input to feedstocks through various combinations of suspension HVOF spray parameters. Three types of coatings with different particle melting conditions (12%, 51%, 81%) were successfully prepared with dense structures and without visible pores and cracks.

The following conclusions allow us to better understand how the existence of different nanostructures affects the coating properties, providing guidance on coating design for different applications:

- Increasing extents of feedstock melting corresponded to increased rutile contents in the coatings, which led to an increase in overall hardness with a reduced plasticity.
- The as-sprayed surface roughness did not seem play an important role for tribological performance of the developed coatings when their Ra varied between 0.53 and 1.18 μm.

- The coating composed of most agglomerate particles (12% melted particles) had the lowest coefficient of friction, whereas the coating composed of mostly melted particles (81% melted particles) presented the highest coefficient of friction. Results also indicate that a higher fraction of agglomerated particles (proportional to anatase content) were beneficial to the formation of tribo-film at sliding surfaces.

- Wear resistance of the coatings were proven to be not rational to their hardness. The coating with mostly agglomerate particles (12% melted splats) had the lowest wear rate and the coating with moderate melted particles (51%) had the worst performance against wear.

Author Contributions: Conceptualization, F.Z., S.W., H.L.d.V.L. and R.J.K.W.; Methodology F.Z., S.W., B.W.R. and H.L.d.V.L.; Validation, S.W., H.L.d.V.L. and R.J.K.W.; Formal Analysis, F.Z.; Investigation, S.W.; Resources, B.W.R. and H.L.d.V.L.; Data Curation, F.Z. and S.W.; Writing—Original Draft Preparation, F.Z.; Writing—Review and Editing, S.W., H.L.d.V.L. and R.J.K.W.; Visualization, F.Z.; Supervision, S.W., H.L.d.V.L. and R.J.K.W.; Project Administration, S.W. and H.L.d.V.L.

Funding: This work was supported by University of Southampton, the TWI Limited (Cambridge, UK) and the China Scholarship Council (CSC).

Acknowledgments: The authors would like to thank Andrew K. Tabecki from the Surface Engineering section at TWI for his assistance with coating deposition.

Conflicts of Interest: The authors declare no conflict of interest.

References

1. Toma, F.-L.; Potthoff, A.; Berger, L.-M.; Leyens, C. Demands, potentials, and economic aspects of thermal spraying with suspensions: A critical review. *J. Therm. Spray Technol.* **2015**, *24*, 1–10. [CrossRef]

2. Lima, R.S.; Marple, B.R. Thermal spray coatings engineered from nanostructured ceramic agglomerated powders for structural, thermal barrier and biomedical applications: A review. *J. Therm. Spray Technol.* **2007**, *16*, 40–63. [CrossRef]

3. Fauchais, P.; Montavon, G.; Lima, R.S.; Marple, B.R. Engineering a new class of thermal spray nano-based microstructures from agglomerated nanostructured particles, suspensions and solutions: An invited review. *J. Phys. D Appl. Phys.* **2011**, *44*, 1–53. [CrossRef]

4. Gadow, R.; Rauch, A.; Rauch, J. Introduction to high velocity suspension flame spraying (HVSFS). *J. Therm. Spray Technol.* **2008**, *17*, 655–661. [CrossRef]

5. Singh, V.P.; Sil, A.; Jayaganthan, R. A study on sliding and erosive wear behaviour of atmospheric plasma sprayed conventional and nanostructured alumina coatings. *Mater. Des.* **2011**, *32*, 584–591. [CrossRef]

6. Li, J.F.; Liao, H.; Wang, X.Y.; Normand, B.; Ji, V.; Ding, C.X.; Coddet, C. Improvement in wear resistance of plasma sprayed yttria stabilized zirconia coating using nanostructured powder. *Tribol. Int.* **2004**, *37*, 77–84. [CrossRef]

7. Ahn, J.; Hwang, B.; Song, E.; Lee, S.; Kim, N. Correlation of microstructure and wear resistance of Al_2O_3-TiO_2 coatings plasma sprayed with nanopowders. *Metall. Mater. Trans. A* **2006**, *37*, 1851–1861. [CrossRef]

8. Lima, R.S.; Moreau, C.; Marple, B.R. HVOF-sprayed coatings engineered from mixtures of nanostructured and submicron Al_2O_3-TiO_2 powders: An enhanced wear performance. *J. Therm. Spray Technol.* **2007**, *16*, 866–872. [CrossRef]

9. Chen, H.; Zhang, Y.; Ding, C. Tribological properties of nanostructured zirconia coatings deposited by plasma spraying. *Wear* **2002**, *253*, 885–893. [CrossRef]

10. Song, E.P.; Ahn, J.; Lee, S.; Kim, N.J. Microstructure and wear resistance of nanostructured Al_2O_3–8wt.% TiO_2 coatings plasma-sprayed with nanopowders. *Surf. Coat. Technol.* **2006**, *201*, 1309–1315. [CrossRef]

11. Jordan, E.H.; Gell, M.; Sohn, Y.H.; Goberman, D.; Shaw, L.; Jiang, S.; Wang, M.; Xiao, T.D.; Wang, Y.; Strutt, P. Fabrication and evaluation of plasma sprayed nanostructured alumina–titania coatings with superior properties. *Mater. Sci. Eng. A* **2001**, *301*, 80–89. [CrossRef]

12. Luo, H.; Goberman, D.; Shaw, L.; Gell, M. Indentation fracture behavior of plasma-sprayed nanostructured Al_2O_3–13wt.%TiO_2 coatings. *Mater. Sci. Eng. A* **2003**, *346*, 237–245. [CrossRef]

13. Darut, G.; Ageorges, H.; Denoirjean, A.; Fauchais, P. Tribological performances of YSZ composite coatings manufactured by suspension plasma spraying. *Surf. Coat. Technol.* **2013**, *217*, 172–180. [CrossRef]

14. Zhang, F.; Robinson, B.W.; de Villiers-Lovelock, H.; Wood, R.J.K.; Wang, S.C. Wettability of hierarchically-textured ceramic coatings produced by suspension HVOF spraying. *J. Mater. Chem. A* **2015**, *3*, 13864–13873. [CrossRef]

15. Zhang, F. Suspension HVOF Sprayed Coatings for Specialised Applications. Ph.D. Thesis, University of Southampton, Southampton, UK, September 2015.

16. Toma, F.L.; Sokolov, D.; Bertrand, G.; Klein, D.; Coddet, C.; Meunier, C. Comparison of the photocatalytic behavior of TiO_2 coatings elaborated by different thermal spraying processes. *J. Therm. Spray Technol.* **2006**, *15*, 576–581. [CrossRef]

17. Borgese, L.; Gelfi, M.; Bontempi, E.; Goudeau, P.; Geandier, G.; Thiaudiere, D.; Depero, L.E. Young modulus and Poisson ratio measurements of TiO_2 thin films deposited with Atomic Layer Deposition. *Surf. Coat. Technol.* **2012**, *206*, 2459–2463. [CrossRef]

18. *ASTM G133-05 Standard Test Method for Linearly Reciprocating Ball-on-Flat Sliding Wear*; ASTM: West Conshohocken, PA, USA, 2010.

19. Mandzy, N.; Grulke, E.; Druffel, T. Breakage of TiO_2 agglomerates in electrostatically stabilized aqueous dispersions. *Powder Technol.* **2005**, *160*, 121–126. [CrossRef]

20. Joud, J.C.; Houmard, M.; Berthomé, G. Surface charges of oxides and wettability: Application to TiO_2–SiO_2 composite films. *Appl. Surf. Sci.* **2013**, *287*, 37–45. [CrossRef]

21. Ghasemi, R.; Shoja-Razavi, R.; Mozafarinia, R.; Jamali, H. Comparison of microstructure and mechanical properties of plasma-sprayed nanostructured and conventional yttria stabilized zirconia thermal barrier coatings. *Ceram. Int.* **2013**, *39*, 8805–8813. [CrossRef]

22. Bannier, E.; Darut, G.; Sanchez, E.; Denoirjean, A.; Bordes, M.C.; Salvador, M.D.; Rayon, E.; Ageorges, H. Microstructure and photocatalytic activity of suspension plasma sprayed TiO_2 coatings on steel and glass substrates. *Surf. Coat. Technol.* **2011**, *206*, 378–386. [CrossRef]

23. Kozerski, S.; Łatka, L.; Pawlowski, L.; Cernuschi, F.; Petit, F.; Pierlot, C.; Podlesak, H.; Laval, J.P. Preliminary study on suspension plasma sprayed ZrO_2 + 0.8wt.% Y_2O_3 coatings. *J. Eur. Ceram. Soc.* **2011**, *31*, 2089–2098. [CrossRef]

24. Herrmann, K. *Hardness Testing: Principles and Applications*; ASM International: Geauga County, OH, USA, 2011.

25. Rayón, E.; Bonache, V.; Salvador, M.D.; Bannier, E.; Sánchez, E.; Denoirjean, A.; Ageorges, H. Nanoindentation study of the mechanical and damage behaviour of suspension plasma sprayed TiO_2 coatings. *Surf. Coat. Technol.* **2012**, *206*, 2655–2660. [CrossRef]

26. Ctibor, P.; Neufuss, K.; Chraska, P. Microstructure and abrasion resistance of plasma sprayed titania coatings. *J. Therm. Spray Technol.* **2006**, *15*, 689–694. [CrossRef]

27. Ma, C.; Wang, S.C.; Wood, R.J.K.; Zekonyte, J.; Luo, Q.; Walsh, F.C. Hardness of porous nanocrystalline Co-Ni electrodeposits. *Met. Mater. Int.* **2013**, *19*, 1187–1192. [CrossRef]

28. Pawlowski, L. *The Science and Engineering of Plasma-Sprayed Coatings*; Wiley: West Sussex, UK, 1995.

29. Leigh, S.H.; Lin, C.K.; Berndt, C.C. Elastic response of thermal spray deposits under indentation tests. *J. Am. Ceram. Soc.* **1997**, *80*, 2093–2099. [CrossRef]

30. Bolelli, G.; Cannillo, V.; Gadow, R.; Killinger, A.; Lusvarghi, L.; Rauch, J. Properties of high velocity suspension flame sprayed (HVSFS) TiO_2 coatings. *Surf. Coat. Technol.* **2009**, *203*, 1722–1732. [CrossRef]

31. Wang, H.F.; Tang, B.; Li, X.Y. Microstructure and wear resistance of N-doped TiO_2 coatings grown on stainless steel by plasma surface alloying technology. *J. Iron. Steel. Res. Int.* **2011**, *18*, 73–78. [CrossRef]

32. Jaworski, R.; Pawlowski, L.; Roudet, F.; Kozerski, S.; Petit, F. Characterization of mechanical properties of suspension plasma sprayed TiO_2 coatings using scratch test. *Surf. Coat. Technol.* **2008**, *202*, 2644–2653. [CrossRef]

33. Lima, R.S.; Marple, B.R. Superior performance of high-velocity oxyfuel-sprayed nanostructured TiO_2 in comparison to air plasma-sprayed conventional Al_2O_3-13TiO_2. *J. Therm. Spray Technol.* **2005**, *14*, 397–404. [CrossRef]

34. Archard, J.F. Contact and rubbing of flat surfaces. *J. Appl. Phys.* **1953**, *24*, 8. [CrossRef]

35. Prchlik, L.; Sampath, S. Effect of the microstructure of thermally sprayed coatings on friction and wear response under lubricated and dry sliding conditions. *Wear* **2007**, *262*, 11–23. [CrossRef]

36. Xie, Y.; Hawthorne, H.M. The damage mechanisms of several plasma-sprayed ceramic coatings in controlled scratching. *Wear* **1999**, *233–235*, 293–305. [CrossRef]

37. Murray, J.W.; Ang, A.S.M.; Pala, Z.; Shaw, E.C.; Hussain, T. Suspension High Velocity Oxy-Fuel (SHVOF)-sprayed alumina coatings: Microstructure, nanoindentation and wear. *J. Therm. Spray Technol.* **2016**, *25*, 1700–1710. [CrossRef]

MDPI

St. Alban-Anlage 66

4052 Basel

Switzerland

Tel. +41 61 683 77 34

Fax +41 61 302 89 18

www.mdpi.com

Coatings Editorial Office

E-mail: coatings@mdpi.com

www.mdpi.com/journal/coatings

www.ingramcontent.com/pod-product-compliance
Lightning Source LLC
Chambersburg PA
CBHW041217220326
41597CB00033BA/6005